The NUFFIELD TRACTOR STORY

Volume Two

The NUFFIELD TRACTOR STORY

Volume Two
Nuffield and Leyland 1963-1982

Anthony Clare

ISBN 978-1-908397-65-2

A catalogue record for this book is available from the British Library

Published by

Old Pond Publishing
Dencora Business Centre, 36 White House Road
Ipswich IP1 5LT United Kingdom

www.oldpond.com

Frontispiece © British Motor Industry Heritage Trust (BMIHT)

Edited by Paul Middleton
Cover design and book layout by Liz Whatling
Printed in China

Author's note

The company involved with the building of the tractors throughout the majority of this book was colloquially known as British Leyland and, as is described in detail elsewhere, was a merger between the Leyland Motor Company and British Motor Holdings Limited. This latter organisation was the successor to the British Motor Corporation. This corporation was a merger between two large automotive companies in 1952, namely Morris Motors and its subsidiaries and the Austin Motor Company. The Morris Motors' side of the business was originally involved in tractor manufacture and its product formed the basis from which most of the tractors described in this book originated.

Some of these organisational changes took place within the period covered by the book, 1963 to 1982, and the tractor operation was successively under the control of the British Motor Corporation, then British Motor Holdings and finally British Leyland. In order to reduce possible confusion the word 'company' will sometimes be used to describe the organisation involved with the tractor operation.

Contents

ACKNOWLEDGEMENTS

When I wrote the first volume some years ago I stated it had taken longer than I had anticipated. In fact, this second volume has taken even longer. The interest of former members of the tractor business of British Leyland has been a source of considerable inspiration and in many ways I feel somewhat fraudulent for telling what is effectively their story. One of their number, Tony Thomas, is well known to readers of *Classic Tractor* magazine and others have given lectures on the subject, principally Simon Evans and Brian Webb. I do not doubt there are others who would like to have had the chance to write at a little length about the history of the company who, for whatever reason, have not been able to do so thus far.

From the era of the Mini I would like to thank John Poulter for his great knowledge of the tractor, including the survivors and the last offerings by Leyland in the smallest tractor range. I would also like to thank Adrian Bentley for providing a considerable amount of background information and for showing me a few of the earliest survivors. He also showed me one of the relatively few remaining examples of the narrow Leyland 154 model that was produced for a brief period in the 1970s.

From the British Leyland organisation I would like to thank John Patterson, who is a great enthusiast for the marque and appears to have an encyclopaedic memory for facts and figures. His very useful résumé provided basic data which has been a great help in tracking the technical development of the tractors. I would like to thank Ray Runciman, a former company demonstrator and ploughing champion, who has proved to have a great interest in Leyland's history. He has only recently retired as managing director of a major former Leyland distributor, Collings of Abbotsley.

I would like to thank two of the ex-graduates for their considerable help in providing information on technical development. Firstly, Tony Moore, who went on from Leyland to JCB and worked on the tractor development known as the Fastrac. Secondly, Simon Evans, who has recently retired from a senior position within JCB and who became nicknamed Synchro Simon for his work developing the Synchro gearbox.

I would also like to thank the aforementioned Tony Thomas and Brian Webb, and also Ron Kettle. All three worked for a long period for Leyland, having started in the early 1960s with Nuffield and BMC. Tony and Ron later worked for Marshall at Gainsborough where Ron followed the tractor through to the end of the marque when it was produced in Scunthorpe. All three have been a major source of knowledge, anecdotes and interesting stories. Tony has written much about the history of the tractor from his own perspective and has recently produced his memoirs in a single volume. I would also like to thank their wives for their very kind hospitality in providing cups of tea, biscuits and even lunches on occasions. They put up with the loss of their husbands for a considerable number of hours on several days when I was eagerly searching for information with which to complete this story.

I would also like to thank Peter Seymour, a long-time member of staff at the Commercial Vehicle Division of the former BMC at Adderley Park, who was a great help, in particular on the earlier stages of the company. He also produced a considerable number of contacts who were more than helpful.

I would like to thank Bob Harris, who came from the Austin part of BMC and provided quite a lot of information about manufacturing at Bathgate. I would also like to thank Barry Parkes and John Barnett, both of whom are former members of the engineering development side at Longbridge in Birmingham. They provided very helpful information regarding the original involvement of Alex Issigonis and the early stages of partnership with Tractor Research at Coventry. Barry Parkes had worked at Harry Ferguson Research for a period of time and had intimate knowledge of the workings of the organisation. I would similarly like to thank Bob Burlington for information from the former Austin side of the business.

I would like to thank Jeff Kitts, another former demonstrator for BMC, who was seconded to Tractor Research for a while and was involved in testing the Mini at locations south of the Midlands. He was involved in sales in the Far East and had some interesting stories to tell about the success, or rather lack of, in getting tractors into some of those distant parts.

I would like to thank Peter Warr, from Totland Bay

on the Isle of Wight, who is the custodian of archive information and items from Harry Ferguson Research. From the same organisation I would like to thank David Buchanan, who was the stepson of Charles Black, an agricultural consultant and a significant player in the development of the Mini. I would like to thank Michael Thorne, from Devon, who showed me the prototype Mini in his collection and who put me on to David Buchanan. He also showed me a prototype produced by the Standard Motor Company.

From the final successors of Nuffield and Leyland I thank Will and David Charnley and their respective offspring for their help and enthusiasm and for letting me photograph some of the items in their fascinating collection of memorabilia. I would like to thank Elizabeth Bryan, who is one of the leading lights of an organisation called Bathgate Once More, which has started a small museum and collection of memorabilia about the Bathgate plant. The group is thus keeping that short but significant period of commercial and tractor production in eastern Scotland in the public eye.

My gratitude is also due to Richard Cole of Cotswold Vintage Tractors who, in a remarkable conversation, told me some interesting facts about his time in the 1960s when he was involved in Mini testing.

Some of the photographs in this volume came from the British Motor Industry Heritage Trust (BMIHT) and I would like to thank Gillian Bardsley, Jan Valentino and Richard Bacchus who work in the archive section of the trust's headquarters at Gaydon, Warwickshire, for their help. They have a colossal archive of photographs showing all aspects of former British Leyland vehicles of all sorts and, while much of this is still being catalogued and archived, what has been found is of considerable interest.

I would like to thank all those who helped publish the book: Liz Whatling for design, Paul Middleton for editing, Sue Gibbard for indexing and Roger Smith for Publishing.

My gratitude is also due to Elizabeth Edwards of itypelive, to whom I entrusted the task of typing this work, and to Tracey Rowe, who is the proprietor of the business and who helped no end in interpreting the differing needs of typing this manuscript.

I would like to thank John and Ann Carter for allowing me to use an annexe at their charming home in the Cotswolds, where I closeted myself away to spend time finally putting the book together.

This and the preceding volume cover a period from 1945 to 1982 and I hope someone will take up the task of writing the story of Marshall ownership and those who subsequently carried on the manufacturing. Even though this period is relatively short it nevertheless makes a fascinating finale to the story, which ends as far as this volume is concerned in 1982.

CHAPTER 1

Background

In order to give some background to the history of the development of the Nuffield tractor and its eventual demise, it is necessary to provide information above and beyond the significant changes that took place during the period which will be covered by this book.

By the late 1960s the British Motor Corporation (BMC), which was producing tractors under the Nuffield name, was getting into serious financial difficulties. The business manufactured a large number of different car models, many of which were simply 'badge engineered' by producing almost identical vehicles with differing colours and badges associated with older car companies. At that time there was considerable brand loyalty attached to individual models as a result of the social attitude towards cars of different marques, whether they were vehicles produced by former luxury manufacturers such as Wolseley and Riley, sporting marques such as MG or the principal large-scale car makers Austin and Morris.

Cars that had been successfully introduced earlier in that era were by now showing their age, but BMC had little in the way of new models in the pipeline and consequently was beginning to lose share to other companies. Principal among these was Ford, whose manufacturing was being arranged into specific sizes and types, a system which has formed the basis of motor production ever since. More importantly, BMC had little in the way of new capital and little prospect of raising enough to develop new models.

In the same era, politics was becoming dominated by increasingly left-wing ideology, which was to some extent a reaction against a long period in the 1950s and early '60s in which right-wing thinking held sway and with it the freedom for commercial companies to manage their affairs with little hindrance from outside influence. Company management skills were of a low order and dominated by senior staff who culturally considered themselves to have natural skills in overseeing large commercial concerns. Little thought was given to providing management training or the application of financial management and other skills that required modern thinking in marketing, sales and personnel management.

In the case of BMC, it became obvious that some sort of state partnership would be necessary for it to survive. The failure of such a large commercial concern, however, would have had serious political consequences for the mainly Labour MPs who represented the industrial towns in which motor and associated engineering manufacture took place. There thus arose an uneasy relationship between BMC and the Labour government of the time. The left-wing approach was to interfere in the running of such private companies, particularly if it was felt that their failure would lead to substantial unemployment in Labour-dominated constituencies. In addition, it provided a degree of satisfaction among left-wingers that such action could be seen to have succeeded over right-wing ideas in the running of commercial enterprises.

By the mid-1960s, the financial position of BMC, which by then had changed its name to British Motor Holdings (BMH), was deteriorating significantly and there was a real need for some outside action to save the business. In the late 1960s, the Labour government set up the Industrial Reorganisation Committee, chaired by Tony Benn, the Minister of Technology. It attempted to resolve the difficulties of the motor industry and forced the merger of BMH and another motor and engineering group, the Leyland Motor Corporation (LMC). Until the merger, LMC had been much more financially successful than BMH. It had a strong involvement in commercial vehicles of all sorts, although it produced far fewer makes of cars than BMH because it owned fewer of the older companies and therefore oversaw fewer different designs.

The Industrial Reorganisation Committee's recommendations led to the creation of the British Leyland Motor Corporation (BLMC). Sir Donald Stokes, who had formerly been in charge of LMC, was appointed chairman. Almost immediately, Sir Donald realised there was a need to rationalise the dominant car business because there were now even more ranges of vehicles which were competing internally with each other. What was obviously required was a standard range produced to meet specific and market requirements of the sort that was being effectively produced by Ford and, to a lesser extent, by other manufacturers such as Vauxhall and the Rootes Group.

Sir Donald Stokes became chairman on the merger with BMC in 1968. (©BMIHT)

manufacturing as these internal changes were those relating to the national economy. In the 1960s, Britain had seen significant change and when Labour came to power in 1964 after a long period out of office many people felt that a new era, not just of social change but in economic and industrial change, had to take place. It needs to be borne in mind that the preceding 20 or so years had been dominated by the enormous impact upon the British economy of its requirement to repay the crippling debts of the Second World War, the funding for which came out of the pockets of the nation.

By the 1960s a large amount of capital was required to modernise industry – not just its machinery and plant but also

The merger encompassed a staggering total of well over 100 companies, many of which were quite small producers of specialist vehicles, but included were larger-scale industrial concerns which were producing components, castings and a variety of other products predominantly for the internal car and commercial vehicle market. All these companies required immediate finance to continue and additional capital if they were to be sustained and grow. The task that faced Sir Donald was all but impossible given the circumstances and this was made far more difficult by problems in the national and world economy, for reasons that will be explained.

The commercial side of the BLMC business included the Bathgate plant in West Lothian, Scotland, a modern factory producing a substantial range of tractors and commercial vehicles. There was not the same need for new capital and reorganisation because the plant was relatively modern but a number of changes were to occur which will be discussed later.

Of almost as much influence on British vehicle

to adapt and change what many regarded as the out-of-date thinking that pervaded much of British industry at the time. Much of continental Europe had the benefit of large amounts of American investment via the post-war Marshall Plan and by the early 1960s was reaping the benefit of new factories, plant and modern ideas. In one sense the enormous destruction of large parts of industrial mainland Europe provided the impetus for large-scale reconstruction. Britain had much less industrial plant damaged and consequently had to carry on production of all sorts with outdated factories, machinery and – many would argue – ideas and management.

There was optimism in 1964 that the Wilson government would begin to provide the much-needed impetus and many would have seen the more recent substantial changes in the motor industry in the form of new plants, such as those at Bathgate, as evidence of the way things could progress. At the time BMH was getting into financial difficulty in the late 1960s, however, the Labour government was beginning to run out of steam. What is more, there was a serious balance of trade deficit, which led to severe pressure on the pound. This in turn led to the need for the Government to request financial assistance from the International Monetary Fund.

This was not a propitious start to the refinancing requirements of the new British Leyland Motor Corporation which, although now given impetus by the Industrial Reorganisation Committee, nevertheless was left with having to find enormous amounts of capital in order to design and introduce new models, particularly for the car part of the business. It also needed to rationalise a large number of plants, which by then amounted to 40 or so spread across the country. That rationalisation would potentially lead to many job losses and create problems for politicians in whose constituencies the plants were situated.

The political consequences of the country's financial difficulties led to a change of government in 1970, when the Conservatives took power under Ted Heath. Today, Heath would be considered quite a left-wing politician but at the time that counted for little when the political situation in the country as a whole had taken a pronounced left-wing step. Consequently, many of the props of the previous administration were left in place and the difficulties the new administration faced were much exacerbated by militancy among the trade union movement. A pattern of damaging strikes developed in the early 1970s and these were to become a characteristic of the decade. A serious dock strike was followed by settlements with a whole range of public workers which increased the financial woes of the country.

The situation was made far worse by the Yom Kippur War between Israel and Arab states led by Egypt and Syria in 1973. This sparked a severe restriction in the supply of oil and precipitated a power and transport crisis the like of which has not been seen since. The national speed limit was reduced to 50 mph and the working week was cut to three days to conserve fuel and power. There were extensive power cuts to conserve energy, and television channels went off air earlier in the evening to cut electricity use.

The situation became far more serious when a dispute arose with the National Union of Mineworkers in 1974. The Conservatives attempted to make political capital out of the situation by calling a General Election, hoping the electorate would back their tough stance in standing up to the miners' demands. However, the Tories were defeated and the return of a further Labour administration gave even more ammunition to the unions.

By now, serious industrial unrest was beginning in the motor industry in particular, with a whole range of wildcat and semi-official strikes. Also at this time the militant union convener Derek Robinson – subsequently nicknamed Red Robbo by the press – came to power at BLMC's Longbridge plant.

All of this had a dramatic effect upon the fortunes of BLMC and the company became effectively bankrupt by 1975. At a time when it was necessary to close plants, reduce over-manning and introduce labour-saving machinery in order to rationalise and become more competitive, BLMC was severely constrained from taking action and financial catastrophe was all but inevitable.

In order to allow the company to survive in some form, the Government asked leading industrialist Sir Don Ryder to produce a report, which was presented in 1975. This recommended streamlining the 'old' – actually only six years old – concern, and the creation of a new holding company, called British Leyland Limited (BL), to take over the business of the previous BLMC. The new British Leyland was partially owned by the government, with the remaining – almost worthless – stock remaining in the hands of private and corporate investors.

The report also led to the reorganisation of the company into four separate divisions. Of these, the Leyland Truck and Bus Division employed about 31,000 people in 12 locations and produced 38,000 trucks, 8,000 buses and about 19,000 tractors per year. Some agricultural machinery was produced in another part of the business, manufactured by Barfords of Belton, which was a small subsidiary of the construction equipment part of the BL business. Its somewhat anomalous situation initially kept it separate from the tractor business and its production was carried out at a separate location in Lincolnshire.

Eventually some form of reorganisation emerged and, with some modern cars being produced, the fortunes of the new business seemed to improve. Significant investment was made at Longbridge in what was to be a long overdue successor for the Mini, which by now had been in production for more than 16 years largely in its original form. Two years later, in 1977, Sir Michael Edwardes was appointed chief executive and further changes took place to the Car Division, which does not concern us as far as the tractor story is concerned.

Further organisation change took place in 1978 when the company formed a new group for its commercial vehicle interests called British Leyland Commercial

Vehicles (BLCV). Included in this new group was the Leyland vehicle business, which manufactured tractors. This group also incorporated a number of the more specialist manufacturing parts of the business. Other smaller specialist manufacturing concerns were not thought to be sufficiently profitable to warrant continuing ownership and many of these were sold to outside parties.

All these difficulties had an impact on the tractor business. The requirement to reorganise car manufacturing had all but financially exhausted the company, which left a shortage of resources, particularly capital, to make the sort of developments in new models on the commercial and tractor side that was thought desirable. Elsewhere, comment is made about why the development of various new components such as engines and transmissions was, to some extent, hindered by the need to minimise the requirement for new capital. It says something for the technical ability of those involved in designing such components that what they produced was as successful as turned out to be the case.

Leyland's main competitors in the tractor market at this time were Massey Ferguson, Ford, David Brown and various small concerns which imported such vehicles. These rivals were less restricted by the ancillary problems of the groups to which they were attached and in some cases, such as Massey Ferguson, were effectively a large-scale concern involved in a single activity. Ford and Massey Ferguson had been the dominant players for many years, producing vastly greater numbers than Leyland and the smaller manufacturers and importers.

No doubt these factors had an impact upon the company's position and when requests were made for new capital towards the end of the 1970s to keep the tractors at the forefront of technical development, it was made very clear there would be little or none available. Indeed, it was becoming increasingly obvious to senior management at that time that without such investment the tractor business had a limited future.

The introduction of the Harvest Gold range with a completely fresh external appearance, and with the new Synchro box and the slightly older but nevertheless satisfactory engine, had produced a tractor of considerable merit. Steps were being considered by the end of the 1970s and beginning of the 1980s to secure a commercial alliance with another tractor manufacturer to allow production to continue. Bob Beresford, a director of the company, stated subsequently that John Deere was being considered as the target for an approach about a possible merger.

This, therefore, is the situation that faced the company by 1982 and, with British Leyland actively divesting itself of ancillary interests, it needed little imagination to see that the relatively small number of tractors being produced and the requirement for future investment made the removal of this part of the business from the group a necessity.

On the national political scene there had been a dramatic change when the electorate decided that the numbers of strikes, together with the ensuing job and financial insecurity, required a radical rethink in the way Britain was governed. This led to the election of a new Conservative government in 1979 under Margaret Thatcher, who was to appear to hold very right-wing political views. It was imperative that she corrected the financial misfortunes of the country and substantially reduce state holdings in industry. Consequently, the Leyland tractor business was quickly identified as being one that should be sold and active steps were taken to find a buyer.

CHAPTER 2

The Mini Tractor and its Successors

From the early days of the development of the medium-sized tractor, which was put into production at the end of 1948, Sir Miles Thomas had made suggestions about a smaller version. Sir Miles, who was at that time vice-chairman of Morris Motors, was also chairman of the Tractor Development Committee. This body was in charge of developing what was to become the Nuffield tractor. This information was contained in correspondence from Sir Miles to Henry Merritt, the tractor engineering designer. Merritt was very dismissive of these attempts by Sir Miles to extend the design remit, largely because he had been struggling with inadequate resources of draughtsmen and associated engineering skills and did not want to dilute the effort being made to design and produce one tractor by having to start work on another.

Merritt was subsequently asked in early 1947 to consider the abilities of an engineering company called Aldic Engineering Limited in Letchworth, Hertfordshire, which Morris appears to have considered purchasing. Aldic had produced a tractor for market garden purposes known as the Gravely. Merritt was pessimistic about this particular venture, perhaps irritated that his attempt to keep the main project going was once again being potentially side-tracked. He concludes in his report about his visit to Aldic as follows: 'There was no need to make these tractors at Letchworth – no need to make them at all!' Little further was heard about the development of the small tractor and by the end of that year Sir Miles had left the company. No doubt by then Merritt had successfully proved that all company resources in terms of design and development should be devoted to the task of producing the original machine.

Little further appears to be recorded about smaller tractors for some years and written records of tractor activities after Sir Miles' departure are in any case very limited. However, it is noticeable that photographs of agricultural shows in the late 1940s and early '50s, reveal that a BMB President tractor was often shown on display alongside Nuffield's own tractor on various dealers' stands. The President used a Morris 10 hp side-valve engine, so there was a tenuous relationship with Nuffield, but it was not particularly successful and by 1956 production had ceased. The business, including a stock of parts, was

subsequently sold to HJ Stockton Ltd of London and the tractor continued in production, albeit under the name Stokold, and used a different air-cooled engine from the Morris side-valve unit.

In the mid-1950s, the Italian Nuffield tractor distributor Vittorio Cantatore of Turin sent one of its

An Italian-built Field Boy seen a few years ago at a tractor restorer in northern Italy.

A further front view.

Field Boy tractors to the United Kingdom. Quite why the tractor was given an English name is difficult to understand unless Cantatore was keen to impress upon BMC it had the resources to build a vehicle the

Note how similar the colour is to the Nuffield orange.

The Field Boy is fitted with a B series 1,500 cc diesel engine.

Engine cover markings. Note the similarity of the wording to the original Nuffield badge.

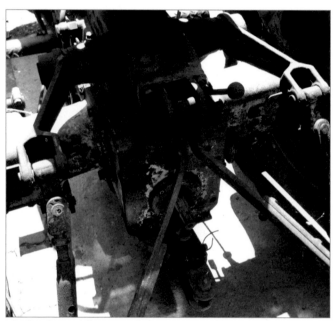

The insubstantial nature of the rear transmission and hydraulics are in marked contrast to the more robust equipment found on the back of contemporary British-built tractors.

company might wish to import. The tractor was presumably made to suit what Cantatore thought was ideal for its local market and possibly someone senior from Nuffield had mentioned to the Italians Nuffield's interest in a smaller model. Cantatore was such an important distributor that Nuffield relations were conducted with the manufacturer at the highest level.

The Field Boy was subsequently tested at Hill Farm, Nether Whitacre, near Coleshill, east of Birmingham, where a number of tractors featuring prototype engineering developments were put to the test. The fact the tests only appeared to extend to field trials perhaps suggest the tractor was not as technically successful as it might have been. The tractor was fitted with a 2.2 BMC diesel engine based upon the B series.

A later model was imported several years

afterwards for further testing. This was probably one of the post-1961 Field Boys, which was updated from the earlier 1950s model. Pictures show this tractor had what was presumably an Italian-manufactured gearbox, rear transmission and hydraulic lift. These components appear to be relatively modest in size, which perhaps suggests the tractor may have been imported on this second occasion for testing for Cantatore's benefit. By this time BMC was committed to the development of its own Mini tractor, which had commenced in 1960.

Prior to this, BMC had become involved with an outside consultancy organisation known as Tractor Research. This was a subsidiary of Harry Ferguson Research Ltd, which had been set up to develop and exploit various patents for a number of automotive component designs. Harry Ferguson Research had been providing testing and development services to BMC for a number of years for engine and other automotive engineering components.

Included among these design concepts was a four-wheel drive technology and this was subsequently put into commercial use under the title of the Ferguson four-wheel drive system, abbreviated to FF and eventually used on the Jensen Interceptor car. In addition, development was being undertaken of a form of clutch that could be used connected to a conventional gearbox to provide similar advantages to a semi-automatic transmission. A form of anti-lock braking and limited slip differential were also being developed and Ferguson was keen to see all these features put to automotive use.

A Field Boy has been imported from Italy and was shown recently in restored form at an East Midlands tractor show.

A completed prototype in what appears to be a red oxide finish used to prevent identification of new models when under research. Note the Ferguson TE20 in the background.

In 1953, Harry Ferguson merged his tractor manufacturing business with that of Massey-Harris of Canada and he resigned from the company a year later. As a result he was barred from further involvement in tractor manufacturing for five years. Near the end of this period Ferguson had started to consider the possibility of further tractor design and had set up Tractor Research to undertake the necessary development work for a TE20-sized machine. Harry Ferguson Research was located in a large industrial building at Toll Bar End on the south-east outskirts of Coventry. Tractor Research was in a building separate from the main automotive development activities. By the late 1950s, Ferguson had produced a multi-page specification for the tractor and was giving active consideration to the use of some of his other automotive technology in the machine. For a variety of reasons, those proposals were not continued because it became apparent that the technology was not suitable for tractor use.

The tractor Ferguson was contemplating was one similar to his TE20, the Grey Fergie, which had ceased production in 1956. An early outline sketch showed the overall appearance of the proposed machine and it was noticeable that the height was less than that of the TE20. Discussions started with Massey Ferguson in 1959 about its possible involvement in production of a further small tractor, which would have been complementary to the TE20's successor, the FE35. This was physically larger and more powerful than the TE20 and Ferguson considered there was a continuing market for something similar to his original design.

Contemporary correspondence from Massey Ferguson indicates outline talks were held between Ferguson and senior executives at the company for well over a year. During those discussions it is interesting to note that Ferguson claimed to have been in talks with other manufacturers. There is likely to be some truth in this because, with the range of his own vehicle component developments, he was no doubt keen to demonstrate these and bring them to the market as quickly as possible. He needed funding to pay for what must have been a relatively expensive research operation, which at that time was probably not earning much money.

Ferguson had held a series of demonstrations for car manufacturers at his home at Abbotswood in the Cotswolds showing the benefits of driving a conventional two-wheel drive car with a four-wheel drive. No doubt BMC would have attended such a demonstration, but it would appear that at about the end of 1959 Harry Ferguson Research approached the company about providing facilities for testing a single-cylinder engine it was developing. These discussions would probably have taken place at Longbridge, where BMC had extensive engineering development and testing facilities and where Alec Issigonis was head of engineering. It was said by a senior engineer involved in engineering development that Issigonis had an active input at the start of the tractor project. It will probably never be known as to who approached whom about the idea of building the new small tractor, but Ferguson was always someone who was keen to put forward his ideas to whoever might wish to hear them. The reader will recall from the first volume that, just after the Second World War, Ferguson demonstrated his proposed new tractor that became the TE20 to Lord Nuffield and Sir Miles Thomas. The idea was not well received by the top Morris men at the time, although Sir Miles was courted by Ferguson some years afterwards for a possible executive role in connection with his other automotive engineering ideas.

The Mini tractor project appears to have begun in earnest by 1960 when Alec Senkowski, who had been a senior development engineer at Massey Ferguson, joined Tractor Research to head the engineering team. He recruited a team of engineers to undertake the necessary design and development work and the project got under way in 1961. The engineers were: Gordon Edwards in charge of transmission design; Bruce Cosh in charge of tinwork design; Ray Tyrer in charge of engine design; Frank Inns in charge of hydraulic design, and Geoff Bunton and Dennis Langton in charge of development.

Senkowski died in 1965 before the fruits of his labour come to the market at the Smithfield Show in December that year. His position was taken by another Harry Ferguson Research employee, John Withers, who stayed in charge until Tractor Research ceased tractor development work for BMC just after the merger with Leyland in the late 1960s.

Each of the engineers had a draughtsman allocated to him and it would appear the design was well advanced in 1961 with the manufacturing of a batch of at least four prototype tractors taking place in late 1962 and into 1963.

Pre-production Mini tractors being assembled by Tractor Research at Toll Bar End.

Two views of pre-production tractors at the Toll Bar End workshop.

Although these tractors were technically very similar to the production models they differed in detail in the fixings for the wheels, the styling of the bonnet and mudguards and, more importantly, the transmission. The prototypes had a gearbox with three principal gears and a two-speed range, giving a total of six forward and two reverse gears. The production tractors had a nine-speed box with the three gears being provided with three separate ratios. In the middle ratio the first and reverse gear were opposite each other, making the matter of forward and aft motion simple for loader-type operations.

The driving position with the two gear levers in primary or secondary. It was a well-thought-out design and made for a particularly compact tractor. Note the seat located on top of the fuel tank.

An alternative layout featuring an under-swept exhaust and equipment to operate a loader. Note the provision of weights on the front wheel, probably not a fixture that one would see much on preserved tractors.

At the heart of the tractor was the A series diesel engine, which at the time was claimed to be the smallest production diesel in the world. Austin had developed the original unit in the early 1950s for car use. It was based upon the same engine used in the Mini car but was fitted with dry liners and had a capacity of 950 cc. The cylinder head had to be completely redesigned to facilitate fuel injection arrangements and this had been developed by Ricardo Engineering, which subsequently acquired Harry Ferguson Research. The cylinder head was based upon a design called the Comet and in addition a strengthened crankshaft had to be provided to cope with the internal stresses. The engine head provided a compression ratio of 22½:1, which was a very considerable amount in its day.

The engine was initially developed at Longbridge and final tuning was carried out at Morris Engines in Coventry under the direction of Eddie Maher and Jack Goffin. An extensive range of tests was carried out, chiefly by Tractor Research at Toll Bar End, but with cold-weather testing undertaken on rigs at Longbridge. The tractor required a considerable amount of power to start, so much so that it was sometimes jokingly said it was the only engine in the

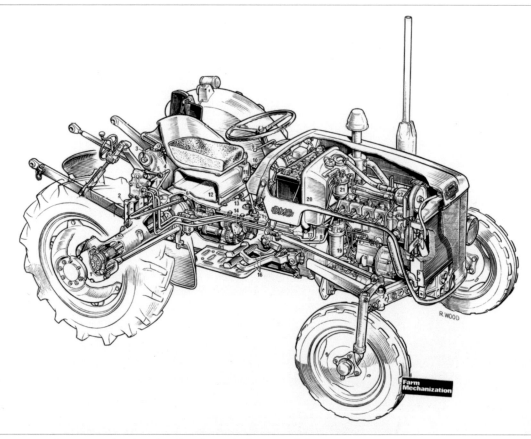

A cutaway drawing showing the main features of the tractor.

A cutaway tractor often used for promotional purposes at shows and exhibitions. (©BMIHT)

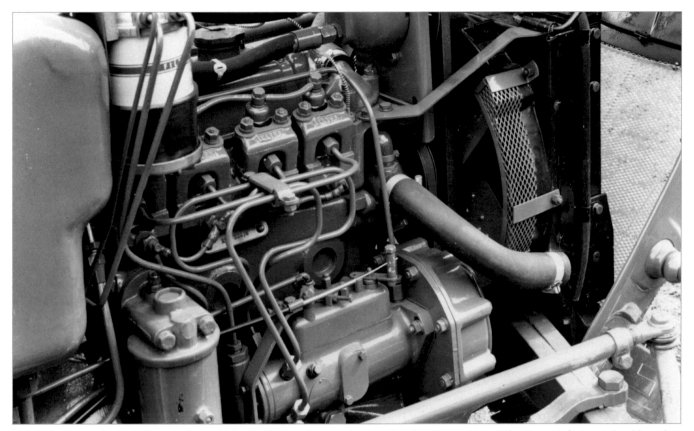

This is the four-cylinder 958 cc engine, which was the smallest production diesel in the country at the time. Unfortunately, although it nominally produced 15 bhp, it was found to achieve less than this under test.

world where the battery was as big as the cylinder block. The first engines had components machined by Rolls-Royce, which were found to be not well made and required in-house machining by BMC to correct. The Tractor Research engineers discovered on a test rig that engine power was only just over 12 bhp, a significant reduction on the 15 bhp claimed by BMC. This led to a counter proposal being put forward to use an alternative engine such as the air-cooled, two-cylinder diesel engine used in the David Brown tool carrier, which had an output of 14 bhp. However, with the amount of effort put in by BMC to develop the diesel version of the A series, the politics of the business dictated there was no possibility of such a change.

The opposite side of the engine. It was jokingly said the engine was so small that it was one of the few in the world where the battery required to start it was almost as big as the powerplant itself.

The rear layout showing the fixed position linkage for the lift arms. Notice the neat way in which the number plate could rotate and be fixed in position for use on the highway.

One of the practical consequences of this power shortfall was that it effectively precluded the use of conventional draft control by means of a hydraulically actuated cylinder opened by forces exerted on the linkage of the tractor. This was because when the cylinder operated to overcome external forces on the implement, a small amount of engine power was absorbed and with the Mini producing so little, this could clearly not be tolerated. Frank Inns, the designer of the hydraulic system, therefore produced a design which became known as 'free linkage' and much of the development took place on a President tractor.

This amounted to a system of preset positions for the lift arms, each one of which provided a small amount of depth control within the fixed limits of the hitch point's operation, five hitch points being provided in all. Thus, any particular implement would have an optimum position depending upon the weight and draft applied. If an implement was being used with a high weight-to-draft ratio then the top position would be used, and for an implement with a

relatively low weight-to-draft ratio, the bottom position would be used. Within the individual range of operation there was only a modest amount of movement but one that would guarantee that the depth of the implement in the ground would stay constant and not impose such an excess of weight transfer on to the rear wheels that the tractor would rise at the front. This was always one of the big selling points made by the Ferguson system and one that was the subject of a model tractor on a board that was used by sales staff to demonstrate its benefits.

It was, of course, possible to lift the implement a modest amount in use and it could be entirely lifted at the end of a row of work but implement adjustments were not, generally speaking, expected to be used on a regular basis. It is interesting to note, however, that in field demonstrations, particularly in areas where ground conditions were poor or where tractors were struggling against an incline, this technique was used by demonstrators to maintain progress with the tractor and prevent stalling.

Tractor Research personnel with a prototype on trial. (©BMIHT)

Extensive testing was carried out on the Mini before its launch in 1965, as will be described in Chapter Six. These were initially carried out on a farm on an estate near Charingworth in the Cotswolds, where a prototype was tested against other tractors acquired by Tractor Research, including a Porsche Junior, a French-built Massey Ferguson 825 and three President/Stokold machines.

The agricultural adviser to Tractor Research was Charles Black, a director of Harry Ferguson Research. He had come to know Ferguson through working for a company called Lenfield in Kent, which was one of the first engineering firms to undertake specialist conversion work for vineyard and hop fields, producing narrow versions of Ferguson TE20 tractors. Black was also a director of a firm called Stockton, which had taken over the assets of President and manufacturing rights to its tractor. Consequently, three machines were procured by Black and utilised for the project. Black was also involved in the import and export of agricultural machinery, particularly to France, and it is likely the use of a French-built Massey Ferguson tractor was directly attributable to this connection. It is probably for similar reasons the Porsche tractor was also tested.

It needs to be understood that the original engineering design and development team probably had relatively little knowledge of agriculture, although Senkowski had worked on Ferguson's earlier tractor projects. They were specialist engineers used for their design and engineering experience but that did not necessarily imply they had any great knowledge of agriculture in general and its need and use of machinery. That input was provided by Black and it is understood that he was instrumental in persuading Midland Industries Limited to produce the hydraulic front loader – the MIL – for the tractor. In addition, a rotavator was developed for the tractor, again it is thought due to the efforts of Black. He produced an amateur film showing the prototype at work, which is of considerable interest as it shows the machine in a variety of locations and performing a number of tasks.

It appears to have achieved most of these without any particular difficulty but it is noticeable the farms concerned have relatively flat terrain and one questions whether the tests would have been quite so successful had the tractor been tested in a more hilly environment.

The Charingworth trials were the forerunner of a more substantial group of tests that were undertaken with direct involvement from BMC and which are also described in Chapter 6. The exact number of tractors made prior to production cannot be determined but it would appear there were at least four prototypes, which had simpler tinwork and differed in features which have been previously described. Later tests at a farm at Enstone in the Cotswolds appear to have utilised machines with production tinwork and one assumes they would be regarded as pre-production models.

The launch of the Mini outside the new sales centre at Longbridge. (©BMIHT)

The range of implements produced or found to be suitable for the Mini was remarkable. Here a two-furrow Salopian Kenneth Hudson plough is at work behind the tractor.

The tractor was announced in a press release on 9 November 1965 and it was launched to the trade at the Longbridge works, where the company had built a striking new sales and display building in concrete and glass, nicknamed the Elephant House. Practical demonstrations were staged at a nearby farm to the west of Birmingham. The public's first official glimpse of the tractor was at the Smithfield Show in December 1965 but despite all the hopes the company had for the new machine, the unfortunate fact was it did not achieve sales as expected. In the preceding year the company had produced about 15,000 of the 10/42 and 10/60 models and it was considered these figures might double when the Mini went on sale.

The sales situation became so serious that the company made the highly unusual decision to relaunch the tractor a year later on a more ambitious scale. This event took place at Alscot Park at Atherstone-on-Stour, south of Stratford-upon-Avon, and a formal invitation and programme of proposed events was produced showing the tractor at work in a wide variety of operations. Everything from ploughing to arable cultivation, seed-bed preparation, row-crop work, turning, baling, trailer work, material handling and fertiliser spreading were carried out in different areas on the same farm in November. This event was supported by a number of implement manufacturers, whose products could be used with the new tractor.

Despite the additional marketing and publicity, sales continued to disappoint. The exact sales figures for the period 1965 to 1968 are unknown but it seems that only a few thousand were sold. In addition, the overseas markets were nothing like as receptive to the tractor as had been hoped. The fundamental problem was that the Mini was underpowered and the company soon realised it would have to evaluate what

Scarcely a year after the introduction of the Mini sales were so disappointing it was decided to stage what effectively was a re-launch at Alscot Park, south of Stratford-upon-Avon. This group of BMC staff are employed in demonstration and associated activities.

could be done to improve performance if the tractor were to sell in significant numbers. The lack of power spoilt the reputation of the tractor to such an extent that it needed to be fitted with a more powerful engine. Furthermore, the lack of live PTO (power take-off) and the unusual form of linkage control were cited as additional reasons for disappointing sales.

It is interesting to note that some of those engaged in field testing before production thought the Mini's potential failing was the lack of power and we have seen that in the early stages of design Tractor Research was proposing to BMC that an alternative engine be resourced because of this issue. However, those suggestions were firmly put down because by now BMC had invested a considerable amount of money and effort to develop the A series diesel engine and had geared up for production with all the attendant costs of mass-production machining. The thought of writing off that investment before costs could be recouped must have seemed a commercial non-starter, apart from other reasons discussed elsewhere.

A more powerful petrol engine was to be made available later but marketed to local authorities and sports clubs where the lower-taxed diesel was not available because of the tractor's non-agricultural use. This model was announced in early July 1967 and the claimed output of this 948 cc engine was 20 bhp at 2,500 rpm with a compression ratio of 7.2:1, providing a maximum torque output of 40 lb.ft at 1,650 rpm. Similar output figures were claimed for a Calor gas version of the tractor, thought likely to be of interest to users where the machine would work inside a building. This required a conversion kit, which was sold for £33 10s (£33.50). A gas tank was required to be fitted on either the left mudguard or the rear frame.

A tractor inside an engineering works, again a further potentially significant market for the Mini although in practice relatively few were sold in comparison with other fields of operation. A gas conversion was available for use in situations where poisonous emissions from diesel or petrol engines would not be permissible. (©BMIHT)

A further development of the Mini was one manufactured for local authority use. It was painted yellow and had square mudguards to facilitate the provision of a cab, a popular feature with such users. These machines were frequently fitted with grassland tyres, as seen here.

The petrol variants were much less common but this is a preserved example of a 4/25 at a South-East vintage vehicle display.

Among the potential optional extras was this cab manufactured by Sirocco. Note its enclosure around the engine, presumably designed for harsh climatic conditions.

A Mini in grounds near the Sisis Manufacturing Company in Essex which produced grounds maintenance equipment. This use had not been considered in the early stage of development and yet the Mini was to be successfully used by local authorities and other such organisations. (©BMIHT)

The total number of Minis sold was not substantial. Over the period for which records are available between 1969 and 1979 – when it was produced as the Nuffield 4/25 and later the Leyland 154 – only about 2,500 were sold in the UK. This represented about ten per cent of total world sales of more than 25,000. The tractor sold almost as well to non-agricultural users as it did to those on the farm, which

perhaps came as a surprise to the company. A considerable number of Minis were adapted or sold with grassland tyres for use by local authorities to maintain public spaces and they were also used by organisations such as sports clubs and schools for grounds maintenance. The grounds maintenance firm of Parker and Sons of Worcester Park, south-west London, was the biggest seller and something like a quarter of all UK sales were made through this one agent, which was located in suburban Surrey rather than a farming area. Parker and Sons produced a small brochure indicating different types of equipment that could be used behind the tractor. Glovers of Stratford-upon-Avon also became major sellers.

One of the fundamental problems was that the new model was introduced at a time when tractors were getting bigger and agricultural machinery development was going in the reverse direction to what one would regard as Mini size. One of the perceived

A Mini being demonstrated within a low barn, the type of farm location where it was felt the tractor would be suitable. (©BMIHT)

Two Minis, one of which was built as a high-clearance version by a firm in High Wycombe and is fitted with narrow rear wheels.

advantages of the Mini was its use in connection with smaller traditional farm buildings but at the time it was introduced larger buildings were becoming available which could, of course, accommodate bigger tractors. By the early 1960s substantial grants were made available by the government to farmers for new farm buildings and, indeed, the author started his training at a firm where this was a significant part of the activities of the surveying department in which he worked. The legacy of that generation of Atcost, Crendon, Marley and other buildings can be seen all over the country.

Furthermore, the other fundamental disadvantage was that, despite the remarkable performance of the Mini, it has to be said that much of the publicity material showed it at work in relatively benign conditions, i.e. on flat land and in light soils. There is little doubt that on hill farms, and indeed it was just such stock farms that were cited as one of the principal target markets, the tractor would have struggled with some of the loads it was expected to haul and implements with which it was to work.

From a financial point of view, the difference in cost between the Mini tractor and some of its larger brothers was not that great. Consequently, because of the greater horsepower and flexibility of larger equipment, in many cases it was difficult to see where the Mini offered any real financial benefit to the farmer, despite the publicity claims. Clearly, much bigger implements could be put behind a tractor of medium size. Yet the marginal difference in cost of 30 to 40 per cent was not a great sum given the limitations to which the Mini would be quickly exposed if it were the sole means of power on the farm. One of the biggest markets for the Mini was the United States, where it sold in significant numbers to hobby farmers and those who typically farmed alongside another profession. This was definitely not the market that was initially envisaged as the reason for producing the tractor.

One of the earliest prototype Minis, now preserved in Devon. Evidence of different paint finishes can be seen. The tractor is otherwise much as originally built.

Perhaps if this examination of the Mini seems a little pessimistic, an interesting postscript might provide some balance. A ploughman who regularly attended ploughing matches in the Sussex area had, for a number of years, used a Mini with a two-furrow Ransomes mounted plough. He achieved a considerable degree of success and was a very strong proponent of the tractor. Indeed, so much so that he also purchased a Massey Ferguson 35 and spent a considerable sum on restoration. He subsequently sold the Massey and believes his Mini was very much its equal for all the jobs which he sought to undertake. Indeed, he considers the drawbar output of the Mini was superior in performance to

The original hydraulic lift. Note that a Ferguson-type top link has been used rather than that specially produced for the Mini.

The engine layout showing the area above the front of the block pump used for the hydraulic system.

A pre-production Mini showing the original squared front of the bonnet.

that of the Massey Ferguson, which is praise indeed.

The ploughman also asserts one needed to spend some time working out how to get the best performance out of the hydraulic linkage and believes that, having done so and spent a little time in adjusting an implement, the tractor would perform most satisfactorily. There is, however, no doubt the tractor was underpowered and the fact that on test the engine failed to produce its nominal output put it at serious disadvantage commercially. Within three years a new engine had been installed.

Fortunately, there was another engine within the BMC range that proved ideal – a 1,500 cc diesel unit based on the B series. This engine had been developed by Austin in the 1940s as an alternative for use in the Jeep. It was subsequently produced in considerable numbers for use in the post-war Austin A40 Devon car and was further significantly developed and later designated the B series when BMC was formed in 1952. The diesel variant was used for several applications including marine use, where it saw, and indeed still sees, use in inland waterway craft, including narrow boats. The 1,498 cc diesel engine provided an output of 24.9 bhp at 2,500 rpm and had a compression ratio of 23:1. A 1,622 cc petrol version was also made available with a similar output. The tractor increased in length by just over 2 inches on account of the engine being larger than its predecessor.

Note the remains of the MIL loader.

A further view of the same tractor showing the extensive amount of restoration required.

A narrow 154 in preservation, although the areas of red paint suggest it might have been a 4/25 with fitments from a 154.

A restored early prototype painted in Ferguson grey. One or two of the early prototypes were finished in this colour to disguise the fact they were a new model.

Lost sales relating to the under-powered engine led to an alternative, the BMC B series 1,500 cc diesel unit, being fitted. The Mini produced a nominal 25 bhp and it is seen here in its new guise when designated the 4/25. This engine was to continue in use until the end of production in 1979.

The announcement of the re-engined tractor was made in a press statement on 22 November 1968, in which it was claimed the switch was the result of research carried out over the previous four years. This appeared to be something of an exaggeration given that the Mini had only gone on sale three years before and it must have taken a few months before the poor sales situation started to become apparent, let alone before the company had decided to do something about the matter. However, one of the original Tractor Research engineers recalls the development brief was expanded and a range of three further tractors was to be introduced alongside the Mini. Tractor Research had sketched out the next in the range as a concept and this was to be larger than the Mini, but utilising the B series engine. A early design drawing for this tractor was produced at about the time that the production drawings for the Mini were being completed.

The 'new' Mini tractor was introduced to the public at the Smithfield Show in December 1968. By now, the merger with Leyland Motors had gone ahead and the tractor was trumpeted as the first to be produced by British Leyland, as the company was now to be known. A very positive statement was made at the time of the launch by the new chairman Sir Donald Stokes about both the tractor and the future of tractor development. Within a few years, however, such statements were to appear somewhat hollow as a result of the difficulties the new company was to experience in the 1970s.

In what was presumably an attempt to dissociate the tractor from its lacklustre predecessor, it was given a new name. It became the Nuffield 4/25, '4' being the number of cylinders and '25' the engine output. It was thus badged in the same manner as the larger tractors, with the former description BMC Mini being removed.

A 4/25 at work with a Ransomes TS90 two-furrow plough.

The merger with Leyland saw tractors re-badged with that company's name and painted blue. The 4/25 became known as the 154 and was identical to its predecessor, other than in colour and a cosmetic change to the front grille.

Those changes were not to last for long because within a year the tractor's external appearance was altered again to coincide with the introduction of the larger 344 and 384 models, which featured a number of significant cosmetic changes over the preceding 3/45 and 4/65 models. The principal and most obvious change was the colour. Gone was the familiar old poppy red livery and instead a mid-blue was applied to the external tinwork, with the engine frame and other details painted a darker blue and the wheels silver.

The B series 1,500cc engine was slightly longer than the previous 948. This lengthened the tractor by about 2 inches.

This colour scheme was applied to the 4/25, which henceforth became known as the 154. The '15' of the designation referred to the number of ccs in hundreds of the engine capacity and the '4' referred to the number of cylinders. Mechanically, the tractor was the same as the previous model but there was a change to the grille with the addition of horizontal bars to match a similar feature on the larger tractors and create a uniform corporate image. The Mini was to remain almost completely unaltered until the end of production in 1979 and the most significant change was the introduction of a narrow model in the early 1970s. This had a reduced overall width of 1.16 m (3 ft 10 in), compared with the 1.60 m (5 ft 3 in) width of the narrow model.

Among the many Mini options was a narrow version and the obvious signs of reduction in width at the front and rear are apparent.

A 154 fitted for highway operation.

A 154 with a cab shown awaiting sale at an auction near Cambridge.

This is reputedly one of the last 154 tractors made. It had been in stock for a while so it was probably contemporary with the manufacture of the early Turkish-built successors.

By 1979, the 154, which was still the basic Mini from 1965 but with the more powerful engine, was beginning to look somewhat dated, and in any case had not sold in anything like the quantities which the company had originally anticipated. Furthermore, Kubota and Iseki, both Japanese manufacturers, were beginning to make significant inroads into the world market for small tractors, selling a more modern machine probably at more competitive prices than those which Leyland was able to charge. There was almost certainly a significant difference in quality because the author has personal experience of dealing with Parkers of Worcester Park, which was at that time the largest seller of the 154 in the UK. Workshop staff complained bitterly in the 1990s about the relatively poor-quality transmission components fitted to the Iseki tractor, which at that time had significantly supplanted the 154 among horticultural and grounds maintenance users. One fitter commented that Iseki transmission components and gears were lucky to have a life of more than 18

months before replacement, whereas the equivalent 154 and earlier tractors manufactured by Leyland had rarely, if at all, suffered from gear failures.

The company at that stage had become closely involved with BMC Turkey, a manufacturing plant that had been set up in that country as a result of a commercial tie between Turkey's two main distributors, which had separately sold Austin and Morris products. It had begun to manufacture a range of its own tractors and at that stage was producing a model of similar size to the 154. BL decided to offer this to the UK market, albeit painted and badged as a Leyland tractor.

Hence, in 1981 the 302 was introduced. This was powered by a 1.8 turbo diesel four-cylinder with an output of 30 bhp. It had a similar arrangement of nine forward and three reverse gears, a two-speed PTO and independent drive disc brakes with fully live hydraulics and swinging draw bar. A Q-cab was fitted to comply with current legislation. The tractor had the front axle set back further than its predecessor to

After the cessation of the British-built variants Leyland sourced Mini tractors from its Turkish plant. This is a finished tractor which was designated the 184 because it used a 1,800 cc engine.

improve its turning capability. However, this made a front-loader difficult to mount, which reduced this advantage to potential users.

This tractor thus incorporated various improvements to make it competitive and one suspects the lower manufacturing cost base in Turkey was probably of benefit. Components were still made at Bathgate and shipped to Turkey, which naturally caused resentment among the Scottish workforce because of the risk to jobs. Staff put pressure on the company not to sell Turkish-built tractors in the UK and it is thought unlikely that many were. By now the company had decided to divest itself of the tractor business and issues relating to promoting a weak-selling tractor would have counted for little given the major problems of the business elsewhere.

An unbadged Turkish-built tractor, part of a private collection in the UK.

CHAPTER 3

The End of the Nuffield Era

A line-up of BMC tractors at Longbridge, with the Mini flanked by a 10/60 and 10/42. (©BMIHT)

With the introduction of the Mini, Nuffield had a range of three tractors and was therefore able to provide a model in each of what at the time were known as the heavy, medium and lightweight sectors of the market. Those terms, however, were not used by the company when referring to its tractors. The heavy and middleweight tractors, respectively the 10/60 and 10/42, were the final developments based on the original design first introduced in 1948. At the time of introduction in 1964, the model 10 was still almost identical in appearance to that introduced 16 years earlier and, with a number of other manufacturers having updated their designs in a shorter time span, the Nuffield was overdue for a comprehensive change of specification and appearance.

A pre-production tractor with annotation mentioning Bathgate on the transmission. This is likely to have been one of the Tractor Research vehicles used for development. Note the absence of decoration on the front grille. This tractor does not have the large letters which were frequently put on the pre-production and early development vehicles.

The wording on the side of the gearbox. It is not known who Mr 'C' was.

The intended development of four new tractors was part of the original remit given to Tractor Research at Coventry, although the model that was to be positioned between the three-cylinder and the Mini tractor was never actually developed. To a large extent its purpose was fulfilled by the provision of the 1,500 cc litre B series engine in place of the original 948 cc Mini engine which, as we have seen, had shortcomings as far as power output was concerned.

The 10/60 and 10/42 were to be the subject of development and an updated version of both, designed by Tractor Research, was introduced at the

Tractor Research continued to be involved in the development of new models. In its workshop is a 4/65. (©BMIHT)

The 4/65 had different layout and styling compared with the older 10-60, which was itself a development of the original M4 from 1948. This styling was not to everyone's taste.

An early 3/45. This model was powered by a continuation of the three-cylinder engine that had been first put into the 'Universal Three' in the late 1950s and produced 45 bhp.

Two tractors under test on a farm at Coleshill in the West Midlands. (©BMIHT)

Royal Show in 1967. These became the 4/65 and 3/45 and represented yet another difference in designation style over the previous range. The first number referred to the number of cylinders and the second referred to the horsepower. In the previous model ranges, '10' referred to the number of forward gears, with the second number showing the nominal horsepower.

The arrangements at BMC were unusual in that nearly all the design and development was being carried out by an outside firm but it must be remembered that tractor design by the manufacturer had been on a modest level for many years. When the tractors were originally made at Ward End in Birmingham, the design staff had numbered only three or four under the leadership of Bill Watton. Most work had involved designing changes, either to implements or to tractor linkage, to make sure the tractor could be used with as many of the tools introduced throughout the late 1950s and early 1960s as possible.

When manufacture was moved to Bathgate, the design office was later absorbed into the BMC design facilities at Longbridge in Birmingham. Tractor design stayed there for several years until the merger with Leyland and the subsequent move of the office, among other functions, to Bathgate from the end of the 1960s. Despite the much greater involvement of Tractor Research, Bill Watton appeared to have played a role, probably as the link man between Tractor Research and BMC. From comments made by former BMC personnel, it would appear that Tractor Research employees were actually located in the Longbridge design office and there was a certain amount of loaning of personnel between one company and the other when demand arose. This did not just occur with tractors but also in engine and associated automotive design.

The tractors introduced in 1967 certainly looked different and were designed with what came to be known as 'Euro' styling. This was presumably chosen because it was a not dissimilar look to that used by a number of competitors who were by then beginning to introduce tractors into much larger markets, albeit produced in only a few locations. Ford was a pre-eminent example where the introduction of its 1000 range tractors in 1964 was to produce a harmonised worldwide design, although manufactured in just three locations.

Similarly Massey Ferguson, and to a lesser extent David Brown and International, was also introducing similar significant changes, so BMC needed to appear to be keeping up with the competition.

A fine view of a 4/65 with the Westbury White Horse, Wiltshire on the hillside. (©BMIHT)

A 4/65 fitted with a Sirocco cab and loader filling a trailer behind a Mini. A lot of the demonstrations by BMC at the time showed that the Mini was expected to be integrated into a farmer's fleet.

Beneath the changed external appearance of the tractor, the engine was the same as had been produced previously and was an uprated version of the OEF and OEG diesels that had been used in the 10/60 and 10/42. By increasing the governed speed to a maximum of 2,200 rpm – a ten per cent increase over the previous models – a slight increase in horsepower was achieved. PTO power was uprated and the 4/65 produced just under 53 hp at a governed 1,800 rpm. Similarly, the 3/45 produced 39 hp.

More fundamental change took place with some of the controls and the layout of the engine compartment. The fuel tank, instead of being located immediately in front of the driver, was now positioned at the front of the tractor. That allowed Nuffield to change the angle of the steering wheel and also provided the opportunity to group the controls and fit an instrument binnacle around the base of the steering wheel in a manner favoured by other manufacturers for a number of years. Similarly, the

A 3/45 being driven by BMC's Ron Kettle at a farm near Coleshill, east of Birmingham.

One of BMC's demonstrator staff, Brian Webb, testing the three-cylinder tractor with a three-furrow reversible plough.

throttle was positioned around the base of the steering column as seen on competitors' models.

Acknowledgement of the increased use of cabs came in the form of angled slopes with a flat horizontal surface to the mudguards to make the fitting of outside manufacturers' designs easier and more weather-tight. Furthermore, the handbrake was relocated closer to the driver and a new cushioned seat was installed instead of the previous, rather crude, metal bucket type which just had a rubber bung for comfort. The bonnet incorporated side access panels covering the engine. Gear changing was now via a remotely controlled lever and the high and low ratio lever was moved close to the main lever for easier adjustment.

The larger 4/65 on cultivation duties.

The driving position of a 4/65 showing the more comfortable seat compared with the simpler pressed tin accommodation available on earlier models. By now a conventional dash had been installed with the throttle around the base of the steering wheel, matching the contemporary practice of other manufacturers.

The hydraulic controls were positioned on a single one-quadrant assembly and again placed in an easier position for access from the seat. The draft control lever incorporated a friction device which allowed for step-less variation of the lever position. In addition, the hydraulics were adjusted slightly to give a higher lift and faster drop than had been the case with the previous model. From the operational point of view, the hydraulics were also adapted to give better quality draft control, albeit of the similar Belleville washer design as that provided on the previous Nuffield models. The 10/60 and 10/42 had only been provided with a simple form of draft control with a single-acting top link. The hydraulics for the 4/65 and 3/45 were augmented with tension control to the top link control spring, providing a double-acting link. The draft control valve had additional capacity provided, which doubled the control valve range from zero to full pressure and provided additional sensitive control. This was not entirely successful because the design was based on regulating oil flow from the hydraulic cylinder rather than regulating the oil flow into it. The additional control was improved later with a further redesign of the system and was introduced during later production of the subsequent 384 and 344 Leyland models. An additional flow control valve was fitted into the draft control pump, which allowed working depth independent of engine speed variation.

Instrumentation included a fuel gauge, an oil and

A 3/45 being tested with a set of Pettit mounted harrows.

A youthful team showing a 4/65 being demonstrated with a five-furrow plough. To the right is Vincent de las Cassas, who had been working with Nuffield tractors since the early1950s.

water temperature gauge with a battery condition indicator, tractometer and warning lights for headlamp high beam. The headlights were grouped in a plastic surround, again an indication of the greater use of this material for vehicle design generally, and this allowed loaders to be fitted without having to reposition headlights from their former position at the side of the radiator surround. The entire tractor had a far more 'modern' appearance and, although appearances are always something of a subjective issue, the somewhat bulbous appearance of the surround to the bonnet was not to my mind an attractive feature. It was generally thought that designers in BMC, not necessarily close to the tractor activities, had approved the new style with too little thought and a better result could and should have been achieved.

The arrangement of mounting the engine within a cast frame continued, albeit of a longer design than that on the previous model. The steering drag link and rod was now located within the frames and not on the outside as before. A further indication of the change to more modern practices was the installation of a no-loss cooling system. This was a modern form of sealed component and was provided with the necessary thermostat to the radiator for rapid warming up of the engine.

Additional front-end weight had been available by bolting a substantial single-cast weight to the front frame casting. This was beyond the ability of most farmers, and certainly the author, to lift and bolt in position so in its place a steel bracket was made available on to which up to eight 'hook-on' weights could be attached to give the necessary ballast for the working conditions. These were considerably easier to manoeuvre and could be lifted on and off the bracket

A preserved tractor at the Harrogate Show a few years ago. Most production models were unlikely to have been in such good condition when leaving the factory.

individually. Front-wheel weights were also available if required. The front axle was redesigned to afford an improved means of adjusting and setting front-wheel widths. Despite the change in appearance the orange colour was retained but with the addition of white wheels, the paintwork of which was described as 'Old English' and similar to that which had been introduced on the Mini.

A considerable number of minor changes were made during the two years this model was produced. Among the more significant was the replacement of the Poly V belt with a conventional standard V belt. This was largely because the sound from worn Poly belts could become very unpleasant as they gave off a very high screech-like note, which was somewhat unnerving.

Further changes took place in the form of stiffening to the cylinder block and the head with changes to pistons, fuel bowls and other minor details. The early

tractors had screwed-on plastic trim and this became an unfortunate casualty of one of the less desirable aspects of the engine, namely the excessive vibration which was a feature that caused problems for users and was the cause of significant complaints. The trim vibrated so much that it broke around the fixings to the bonnet and fell off. This was overcome on later models by the use of applied transfers. The plastic surround to the instrument binnacle similarly suffered from the same problem.

One of the other more significant effects of vibration had been damage to the fuel tank. The relatively thin metal became weakened, causing splits to the bottom of the tank and consequent loss of fuel. The tanks had to be redesigned with a strengthened base to overcome this defect.

Additional strength was given to other tractor components, including gears and the PTO shaft. This was necessitated by the greater stresses being placed

A 4/65 under test with a power rotavator at Maxstoke Hall Farm near Coleshill, which was much used by the company for demonstration and development.

A 4/65 at work with a four-furrow plough and fitted with a Winsam cab. Cabs became an increasingly frequent feature on tractors in this era.

upon the tractors, largely due to the use of the larger implements that were being introduced at the time.

By 1967, the manufacturing problems which had affected the production of engines on the new assembly line at Bathgate had been much reduced. The problem of liner corrosion had been mitigated by the provision of anti-corrosive chemical pellets, which would be inserted into the waterways at the time of manufacture. The coating of cylinder liners with further anti-corrosion protection considerably limited the problems seen before. By 1967, and after five years of work, the labour force had become used to the demands of production and with improved training and quality control many of the problems associated with a lack of assembly skills had been eliminated.

The tractors were also

The Queen visited Bathgate during the production period of the 4/65 and is seen here walking around the assembly line. (©BMIHT)

Bray of Hounslow produced a four-wheel drive conversion, seen being demonstrated with a reversible plough at Maxstoke Hall Farm. (©BMIHT)

available with the option of a considerable number of extras and consequently one could be custom-built for a particular farmer. Indeed, it seems probable that the majority of tractors were assembled in this way because, owing to the large number of optional extras, it would be impossible to pick out a range which would suit a standard customer. Part of the redesign had allowed for the fitting of power steering which, with heavier implements and loaders, had become a necessity for users. For foreign markets some changes were to be provided in any case to take account of differing road conditions overseas. For example, many export tractors were fitted with oversized tyres.

BMC produced a table for the sales force showing the performance of the individual models against those of their competitors. Thus one could see how the Nuffield 4/65 compared against the Massey Ferguson 165 and 178, the Ford 4000 and 5000, the David Brown 990 and 1200, the International B 450 and 634 and equivalent models from John Deere, Zetor and Allis Chalmers. The respective comparisons in terms of engine output, PTO performance, belt pulley performance, transmission, road speeds, reverse performance, hydraulics, capacities and dimensions were set out in tabulated form. These documents indicate BMC had produced a tractor which was among the best compared with its major competitors.

Another Bray conversion hauling heavy-duty tillage equipment on a steep hillside.

A lorry carrying a Mini, 3/45 and 4/65, presumably to a demonstration. The company made extensive use of the then nationalised British Road Services for tractor transport. (©BMIHT)

This 3/45 at a vintage vehicle show in the South-East is in a condition in which tractors are more commonly found today.

Some preserved tractors retain their previous equipment, such as this mounted winch at a Hereford vintage vehicle show.

Despite these changes and the hoped-for improvement in sales the new model was not particularly successful commercially because of its perceived problems and unreliability. Tractor Research set about designing a replacement with further external cosmetic changes to the appearance. This replacement came relatively soon after in 1969. That was to be under a different guise and thus it was that the 3/45 and 4/65 ended the Nuffield production era which had started at the end of 1948 and lasted until 1969.

Its successor also saw the end of Tractor Research's involvement with tractor design for BMC, a move regarded by many insiders as a mistake. A large number of the tractors were left unsold when the change of models occurred and BMC attempted to sell these into export markets, presumably at a discounted price, to help remedy its dire financial position.

Not an attempt to put a 4/65 into space but cultivating rice in the Far East.

Given the problems of keeping the front end on the ground, note the substantial weights and use of wheel strakes to improve adhesion.

CHAPTER 4

The Leyland Era

Up until 1968 tractors had been built by a subsidiary of the former British Motor Corporation, which had been formed in 1952 by the amalgamation of the Austin Motor Company and Morris Motors Limited and their various subsidiary interests. The merger was not thought to have been particularly successful by those in the automotive industry. Austin had not produced tractors for many years when the merger took place and consequently the only machines that were produced by the new combined company were those which were described in the previous volume by the Morris Motors subsidiary of the business now known as BMC. The merger with Leyland led to tractors being produced through the combined efforts of the new company.

The commercial vehicle side of the BMC business, involving as it did the Bathgate plant with its range of commercial vehicles and tractors, was not subject to the same imperative need for reorganisation and investment in buildings and plant as other parts of the combined business. Nevertheless, a number of changes emanated from the merger, among which was the introduction of a new design office at the plant.

British Leyland chairman Sir Donald Stokes took a keen interest in the tractor business in the early stages of the merger and made a number of important organisational changes shortly after it. The services of Tractor Research were terminated and tractor design was placed under the control of Ralph Wigginton. After a short spell at Longbridge in the existing office, design was relocated to the Bathgate factory where new accommodation was found for what was to be an enlarged tractor team in offices at the end of the truck assembly building. Wigginton had come to the company from Standard Triumph, where it is thought

When Leyland merged with BMC it had no subsidiary business that had specific knowledge of tractors other than manufacturing on behalf of Ferguson. It had, however, produced an experimental tractor and this is one of the six prototypes.

A detail of the four-cylinder engine used in the experimental prototype. Note the use of what appears to be a paper element air filter.

he had been involved in the design of the Herald car and also had some involvement with the Standard tractor. It seems probable that Sir Donald wanted men around him who he knew would have been imbued with the automotive ethos of Leyland, which he had developed, rather than unknown personnel from BMC.

The Standard Tractor had been designed after the merger between Massey-Harris and Harry Ferguson when it seemed probable the Standard Motor Company's manufacturing arrangements with Massey Ferguson might come to an end. Standard had secretly designed and built a small batch of experimental and prototype tractors which were similar in size to the contemporary Ferguson FE35 and TE20 series. They had additional capabilities built into both the gearbox and transmission to allow the eventual use of larger engines without the need to undertake expensive changes to the transmission components. These prototypes were never put into production but at least two have survived. Indeed, one survivor went to the

Bathgate factory and was kept for a number of years at the Mosside Farm training school. This is believed to be the tractor now in the Crawford Collection at Frithville in Lincolnshire. A further example exists in a private tractor collection in the West Country.

Bill Watton, who had previously been in charge of tractor design in the Nuffield period of production at Ward End in Birmingham, was made an assistant to Wigginton at Longbridge, although he did not subsequently move to Bathgate. There appears to have been considerable reluctance to move from Birmingham on the part of many personnel, although by then Watton was likely to be nearing retirement. The transfer of commercial and agricultural vehicle design to Scotland was not popular among those previously employed in the extensive design office at Longbridge. Subsequently, the post of Wigginton's assistant was taken by Mike Barnes, who had previously worked at David Brown on, among other things, the Selectamatic

Note the simple rear dash arrangements. One assumes this would have been developed further before production.

transmission design. Further recruitment continued with Terry Stephenson from International Harvester, who was involved in clutch, sheet metal and brake design, and Phil Young from David Brown. Three talented former graduates were also taken on: Simon Evans, who designed the new Synchro transmission; Bruce Davies, who redesigned the hydraulics; and Tony Moore, in charge of field testing. It is clear that design leadership was now in the hands of someone whose experience would have been significantly less than that available through Tractor Research.

Tractors were always a smaller part of the BMC commercial activity based at Bathgate but after the merger with Leyland, with its strength in larger commercial vehicle production situated mainly in Lancashire, that position was to become more pronounced. Tractors were now an even smaller enterprise among the enlarged company's commercial vehicle activities. Consequently, finding capital for significant investment in tractor development was to become more of a struggle for the next 13 years. The continuation of tractor sales, however, aided the commercial vehicle part of the business by increasing

demand for engine and transmission components, which reduced the unit costs for the commercial vehicle side of the business.

In the latter period of 4/65 and 3/45 tractor production, development continued to eliminate design weaknesses that farmers had identified as important and which the company had deduced was a factor in the relatively poor sales. The newly merged company offered the chance to produce a tractor different in style and with limited improvements which, together with those previously incorporated, were hoped to increase sales and improve reliability.

Before replacement tractors were introduced, it is interesting to note that in 1968 Keith Sinnott, in charge of tractor marketing, had written to John Withers, who oversaw tractor development at Tractor Research in Coventry, with a view to the latter developing a new 100 hp (actually 98 hp) machine. This would have been something of considerable size and power, well in excess of contemporary tractors produced both at Leyland and elsewhere. It would appear, however, that the proposal went no further, almost certainly because the merger of the BMC and Leyland automotive interests was being completed at the same time with the resulting termination of design arrangements with Tractor Research.

In 1969, two new models called the 384 and 344 – the last in which Tractor Research had an input – were introduced, initially to the trade at an exhibition at Alexandra Palace in north London. A stunt was set up to capture the interest of the assembled trade in which one of the new models burst through a paper wall separating the sales area and the rear of the exhibition space (see page 121). The tractor was subsequently introduced to the farming press and the agricultural community at the December Smithfield Show. Included in the line-up was the Leyland 154 and it has been discussed elsewhere how this was incorporated within the new Leyland family of tractors.

A number of technical changes were introduced on the 384 and 344. The first two numbers of the title referred to the size of the engine in hundreds of cc capacity, the last figure the number of cylinders. Consequently, the smaller tractor, a 1.5 litre unit with four cylinders, became designated the 154.

The larger 384 continued

The first tractors produced by Leyland were developments of what had been previously manufactured, with the exception of the three-cylinder variant.

An early 384 under test. This model succeeded the 4/65.

with the 3.8 litre engine from the previous 4/65 but had uprated injectors to produce an output of 70 bhp, a modest increase on the old model. The smaller 344 used the original 3.4 litre block of the predecessor to the 3.8 litre tractor, which had a bore of 100 mm diameter, the earlier engine dating from the 1950s OE series having a 95 mm diameter bore.

The three-cylinder engine had continued to be produced at the former Morris Commercial Motors works at Adderley Park in Birmingham but was now no longer available. This was not the case with the four- and six-cylinder OE engines, production of which had been transferred to the Bathgate plant when it opened in the early 1960s. The three-cylinder engine was terminated because it was only used in tractors and the limited demand made it uneconomic

The smaller tractor was designated the 344 and actually used the older BMC diesel four-cylinder engine. It is seen here with oversized tyres, possibly an export model.

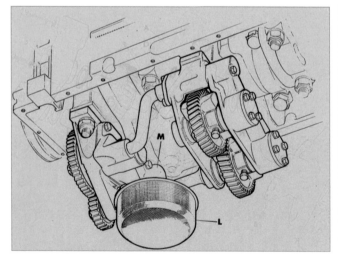

The engine balancer fitted to both four-cylinder tractors. The drawing shows the cylinder block and highlights the engagement of gears with the associated mechanism that evened out the vibration stresses. This occurred during the passage of the piston from the top to the bottom of the cylinder.

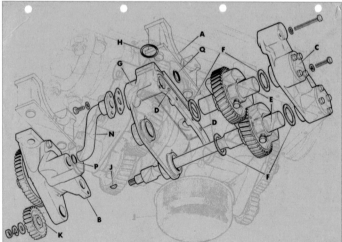

A sketch of the device split into components.

A 344 under test ploughing at an unknown location. By the time the 384 and 344 were introduced most testing had been moved from the West Midlands to Scotland.

to transfer the production machinery. The tractor had a solid frame, which could handle the torsional stresses induced by the crankshaft, but this was not the case in the lighter-framed smaller commercial vehicles in which it had been tried.

The 344's engine produced 55 bhp and both the two larger tractors had a maximum engine output at a governed speed of 2,200 rpm. The increase in crankshaft rotational speed had caused significant problems and the difficulties of engine imbalance and the associated vibration in the four-cylinder models was eventually solved by the provision of what is known as a crankshaft balancer. The imbalance was caused because the centre of gravity in the engine moves in relation to the up-and-down movement of the four pistons. The balancer was fitted into the crankshaft cavity and a gear and counterweight from the balancer meshed with another gear and counterweight on the crankshaft. This smoothed out the irregularities set up at higher engine speeds and reduced the problem of vibration. It was possible to balance one of the old threepenny coins, which had twelve sides, on the bonnet of the tractor without the coin falling off.

The most striking change was the external appearance of the tractor. The familiar Nuffield orange had been replaced by a two-tone blue livery with the lighter blue being applied to the tinwork, such as the bonnet, radiator surround, mudguards and cab where fitted, with engine and transmission components finished in a darker blue. The wheels were painted silver. The tractors were no longer labelled Nuffield but Leyland, although curiously the name Nuffield was applied in small letters beneath the individual model designations on the lower part of the rear bonnet. They were to be the last to carry the Nuffield name.

These tractors had frames of extended length but

The young woman on the 384 was a secretary to one of the senior managers at Bathgate. She said subsequently that she was concerned about the weather as rain was in the air.

had to retain the same wheelbase to make them manoeuvrable enough to fit the normal gate widths and buildings in use on British farms. The styling was rather more square with a sloping bonnet front. Gone was the bulbous-fronted 'Euro' styling of the previous model.

Further technical changes took place in the form of a new cyclonic paper element air cleaner which was positioned at the front of the tractor, as was the battery which was located in a lower portion of the extended frame in front of the radiator and beneath the fuel tank. This made it easier to change the battery, although access problems remained where loaders and other front-mounted implements were fitted. The position was considered better in preventing battery overheating and also provided additional weight at the front of the tractor, which was assisted by raising the fuel tank slightly from its previous location in the 4/65 model. The gearbox, rear transmission and hydraulics were little altered from the previous model except for some minor changes and development.

Before the tractors were exhibited to the trade and the farming community, a publicity field trial took place under the supervision of the local branch of the National Farmers' Union (NFU) on a site at Ratho in Midlothian.

The hydraulics were subject to redesign for both the 384 and 344 and incorporated what was considered to be a relatively substantial and strong hydraulic unit. This was to include position control hydraulics thought necessary to cater for the use of the heavier implements, in particular ploughs and cultivating machinery, that were being developed by

A revised layout of the front of the tractor was introduced. The open bonnet reveals the tank positioned in front of the engine, with the radiator between the tank and the engine.

A number of details were improved to make the tractors more compatible with modern implements. This shows the pick-up hitch with various features and fitments provided for different types of agricultural operation.

the agricultural equipment trade. Controls provided included an up-and-down lever, with a second lever controlling draft or position control and a third providing control to an external service point.

The Ratho ploughing test showed the proposed means of providing the position control hydraulics were far too

unreliable and potentially dangerous given the enormous amount of heat that built up in the rear transmission. Each day a new hydraulic unit was shipped from the factory and was fitted to the larger tractor to overcome these defects. For those reasons the tests were not as continuous as the publicity material subsequently suggested. Consequently, the hydraulics went back to the design used in the 4/65 and 3/45 ranges. Wigginton's assistant, Mike Barnes, believed the hydraulics were never adequately developed and it was not until the Marshall era that an adequate system was to be available – beyond the end of the British Leyland tractor era. Using a mounted mower, hay windrower or fertiliser spreader required the use of auxiliary chains fitted from the implement to the top link anchor bracket. Leyland was one of the few tractor manufacturers at this time that did not incorporate position control into the hydraulics.

In 1970, a new cab was introduced when legislation required the fitting of new safety types in the UK. Tractors for the home market were now provided with what hitherto had been an

The driving position showing the binnacle with instruments.

Leyland was keen to promote the use of the small tractor whenever it felt it appropriate. Here a 154 is loading a trailer behind one of the new 384s. It seems unlikely that many farmers would have purchased a small tractor simply for this purpose.

A preserved 344 at a vintage vehicle show.

accessory. Cabs had been used from a number of different manufacturers up until that date, and indeed Winsam provided those fitted to the two tractors that undertook the Scottish NFU test.

At the Smithfield Show in December 1971, another tractor labelled the 253 was announced. This was a 'new' model designed to appeal in the medium-sized sector and was something which Leyland perceived was not otherwise provided for by the 344, although the 47 bhp output of the new type was not much less than that of the 55 bhp output of that model.

The 384 and 344 tractors used the four-cylinder engines from the OE series, the three-cylinder OEF no longer being available. An alternative three-cylinder engine was available and was purchased from Perkins

The hydraulic layout at the back of the new tractors, indicating a pick-up hitch in position.

The smaller 344 being tested with a McConnel digging power arm on rather difficult boggy land near Bathgate.

The opportunity was taken in 1971 to introduce a further smaller tractor and this was a three-cylinder machine powered by a Perkins engine. The powerplant did not fit the frame of the tractor and consequently a section of frame had to be removed to accommodate the rear of the crankcase. It is seen at Mosside Farm.

A 253 at work with disc harrows.

A further view of a 253 at the rear of the Mosside Farm complex. One of the company's ploughs used for testing and demonstration purposes is in the foreground.

A tractor being used for mowing. It is interesting to note it being used with the older type of cutter-bar mower.

of Peterborough. An earlier variant of the same engine had been used extensively by rival tractor manufacturers for a number of years. This engine would not fit the standard Leyland frame and consequently had to be adapted with a partial side extension frame. The engine and transmission were therefore of integral construction, forming the structure of the tractor. The frame was provided for supplementary purposes to support the battery holder, fuel tank and associated equipment at the front such as weights.

One of the consequences of this change of design was that the fixings between the bell housing of the engine and the transmission provided insufficient rigidity without the previous continuous sub-frame – a number of tractors 'broke' at the engine gearbox joint in spectacular fashion.

The transmission was based on the original 5 x 2 speed gearbox that had been used for many years by Leyland and the rear transmission was a further development of the existing unit, which incorporated the hydraulics system.

Contemporary with the introduction of the 253 was the development of a new range of engines. It needs to be borne in mind that, by 1971, the engines in use were still based upon the original early 1950s OE

units which had been introduced for tractor use in 1954. They were by then nearly 20 years old and becoming outdated.

Comments have been made, both in this and the previous volume, about the inherent problems associated with the enlargement of the original engines from a 95 mm to a 100 mm bore within the cylinder block. Although many of the earlier problems from the 1960s, particularly liner corrosion and head gasket failure, were largely eliminated within a few years, it did not get over the fact that a more modern engine was perceived to be necessary. This was the case not just for the tractors but, probably more importantly, for commercial vehicles as well.

Among the reasons cited by Leyland for development of the new range were that the current engines were perceived as being unreliable with, by that stage, a low top engine speed lacking in power. The 3.8 litre engine which was commonly used in commercial vehicles was, of course, used in tractors but not the larger 5.1 and 5.7 litre six-cylinder units. The new range of engines was designated 4-98, with 6-98 the larger unit. The initial number designated the number of cylinders and the last two numbers the diameter of the cylinder bore. The 4-98 NV engine replaced the previous 3.8 litre unit but both retained a capacity of 3.77 litres. The 6-98 NV replaced the 5.7 unit and the 6-98 DV replaced the 5.1 litre unit. The D sub-designation denoted the de-rating achieved by the use of a different fuel pump to that on the NV model, producing a lower power output.

The company claimed the common bore provided benefits in standardisation of engine components but that would not have been of benefit to many in the agricultural field where farmers held little stock on the farms for major repair. In contrast, large commercial fleet operators with central workshops would have held reasonable quantities of spares.

Apart from the standardisation of the bore size, the block was enlarged by lengthening at the base to extend the stroke to 125 mm from 120 mm. This also provided the benefit of an additional liner seal at the base. The liner was increased in thickness with the decrease in size of the bore, and a revised seating at the top of the liner strengthened the joint and the top

A detail of one of the 98 series engines, actually the six-cylinder version used in the high horsepower tractors.

A re-badged 253 was introduced in 1972 as the 245.

of the block itself where cracking had previously often occurred. The cylinder head was stiffened and new inlet and exhaust valves fitted. The internal water passages were also improved to aid cooling. The injector angle into the head was changed to improve fuel circulation and the injector liner modified. The pistons had a change in the top cavity from a hemispherical shape to a toroidal shape. This was thought to improve gas circulation, so aiding economy and power. Pistons also had a revised arrangement of rings. The connecting rod required modifications to allow for the changes in the stroke length, and revisions to the fixings were incorporated to overcome a previous minor problem that had arisen at the point of fixing to the big-end bolts.

The 270 was introduced to replace the 384. It is seen at Mosside Farm, unusually with rear agricultural tyres and front grassland tyres.

Additional changes took place to the cooling system with the passage from the thermostat housing altered to provide a smoother flow and larger diameter, increasing the flow of water and aiding cooling, particularly in the initial phase of engine start-up before the thermostat opened. This led to a reduction in fan size, which in turn aided heat loss and also reduced marginally the power required to rotate the fan.

Engine output increased at greater revs although a reduced maximum of 2,200 rpm continued for use with the new engines in tractors. However, a better torque output was produced which was a significant benefit. The new engines were claimed to be more fuel efficient and the company saw their benefits as offering more power, speed, torque and pulling ability together with better economy, less smoking, greater strength, longer life and easier and reduced maintenance. This was undoubtedly an engine better able to compete with other manufacturers' products and was technically considered to be one of the best, if not the best, tractor engine available at the time.

The engine, however, did not look greatly different from the exterior, the company looking for every advantage in reducing the cost of replacement castings

by making minimal changes to patterns for cylinder block and head castings. The installation of the new engines into the tractor range began with two new models in 1972 which were designated the 270 and the 255. There was thus another change of designation system from the previous 384 and 344 models. Under the new one the first digit represented the number of driven wheels, the second two numbers indicated engine horsepower. The existing 253 model continued unchanged but was now redesignated 245 in line with the system. The 154 also continued, but curiously the designation was not changed to 225.

The reduced horsepower output of the smaller-output tractor, the 245, was achieved by the use of the 4-98 DT engine with a different fuel pump, a CAV DPA pump as opposed to the Simms Minimec on the larger-output tractors. The designation letter T indicated a tractor engine as opposed to one running at higher revs and used for trucks. Both engines continued to use the harmonic balancer introduced a few years before.

The transmissions, final drives and hydraulic systems were carried over from the 384 and 344 range and, with engine output nominally the same, the updated models did not differ as much as might have

A further model, the 255, was introduced in 1972 and is seen with a front-mounted cultivator and rear harrows.

been thought. The hydraulic system by now incorporated a form of position control, which had been intended for the preceding range. Little was made of the introduction of this feature, which was well behind the introduction of similar systems in rivals' ranges. Ford had introduced a similar feature seven or so years earlier on the Super Major.

The new engines were, in reality, probably more of a benefit to the commercial vehicle market than they were to the agricultural field. A more significant change from the agricultural viewpoint was the introduction of 100 and 85 hp tractors using the 6-98 NT engine, the lower powered machine featuring a DT de-rated model similar to that described for the 255 and 270. It was said the Australian market in particular wanted more powerful tractors. One can envisage in a country where there are large open-scale farms that greater use could be made of machines of that size where implements could be larger than those in more traditional small-scale farming of the United Kingdom at that time. The previous OE engine range contained a six-cylinder engine but this had not been previously used in tractors, although it had been installed by dealerships providing specialist services and by farmers themselves. The high-horsepower tractors were only produced in small numbers until the development of the 4-98NT turbocharged engine

This 245's frame only provided basic shelter. This was presumably available for markets where only simple weather protection was required.

In the same launch period, December 1972, the 6-98 engine was used for the first time. Here it is fitted into an 85 hp, two-wheel drive tractor.

The more powerful variant was the 100 hp tractor, seen at Bathgate.

A 100 hp tractor at work with four-furrow reversible plough. Hauling a plough on an incline of this sort would have required a machine of significant power.

A four-wheel drive, six-cylinder engine tractor preserved in a private collection.

became available and was introduced in the later 282 and 482 models.

A four-wheel drive variant offering the same two engines from the 6-98 range was introduced during this period. These produced respectively 85 and 100 bhp. The four-wheel drive was provided by using a County-manufactured front and rear axle, bought in as completed components from the manufacturer at Fleet in Hampshire. The wheels were of equal size, which was a standard feature of the company's four-wheel drive vehicles.

A further view of one of the earlier four-wheel drive variants.

This tractor seen at an auction will need some attention before it is as well-equipped as the previous example.

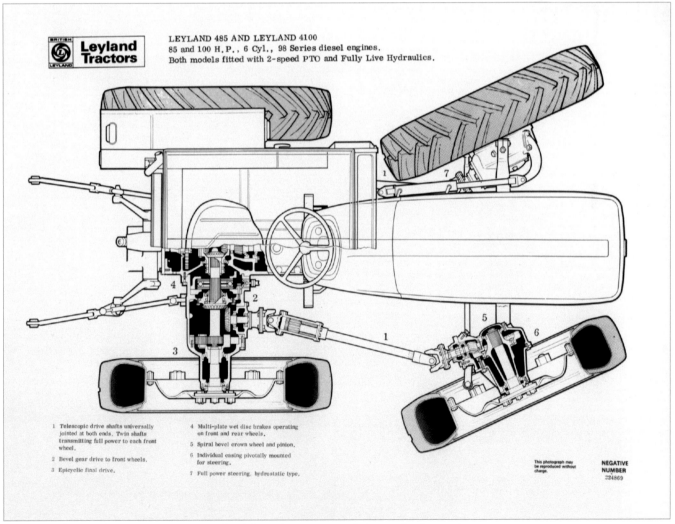

Leyland Tractors
BRITISH LEYLAND

LEYLAND 485 AND LEYLAND 4100
85 and 100 H.P., 6 Cyl., 98 Series diesel engines.
Both models fitted with 2-speed PTO and Fully Live Hydraulics.

1 Telescopic drive shafts universally jointed at both ends. Twin shafts transmitting full power to each front wheel.

2 Bevel gear drive to front wheels.

3 Epicyclic final drive.

4 Multi-plate wet disc brakes operating on front and rear wheels.

5 Spiral bevel crown wheel and pinion.

6 Individual casing pivotally mounted for steering.

7 Full power steering, hydrostatic type.

This photograph may be reproduced without charge.

NEGATIVE NUMBER
224869

A drawing showing how power was transmitted from the rear transmission to the front axle.

The four-wheel drive variants were produced with the use of equipment from County Commercial Cars Ltd. This was available as either a 100 or 85 hp tractor.

A four-wheel drive tractor at work with a substantial plough.

Ray Runciman, one of Leyland's demonstrators, won a ploughing match in the Eastern Counties and here he is showing off his winner's cup with other colleagues.

Instead of steering from the slewing brake, which had been a typical feature of County four-wheel drive up to that date, a conventional steering wheel was fitted. The forward drive to the front wheels was provided from a rear differentially driven bevel gear to two telescopic shafts with universal joints at each end, driving through bevel wheels and pinions to the front wheels. Power-assisted steering was provided as a standard feature. The engine was rubber-mounted and the side frames could fail under load, leading to County stating it would not guarantee the four-wheel drive system and the model was dropped.

The same provision of hydraulics was offered with four-wheel drives as with the two-wheel drive tractors. Production at Bathgate was significant during the period these tractors were available, but the four-wheel drive machines do not appear to have sold in large numbers and the range was discontinued in 1976.

In 1975, three years later after the introduction of the 98 series engines, evolutionary development of the engine to produce more power led to a further facelift of the tractor. Opportunity was taken this time for further changes to the transmission and hydraulic system.

The engine change provided a modest extra amount of power which was achieved by alterations to provide a new lift pump with a two-speed advance control, which gave more accurate governing and better torque characteristics. Stronger axle shafts were now fitted in order to transmit torque to the wheels safely.

The PTO was redesigned to provide smoother engagement and, it was claimed, a longer service life. The hydraulic system was improved with a new design of valve chest installed to cope with ever increasing implement weights. An optional 11/46 final

The 272H featured revised gearing to the rear pinion shaft and had a potential maximum speed of 25 mph where this was the legal limit for agricultural vehicles.

drive pinion could be fitted, allowing an increase in cultivation speed and giving a top road speed of just under 25 mph, the legal limit for agricultural vehicles. In the UK, the limit was lower than in other countries and Bathgate produced a lock to prevent engagement of the top gear. This model was designated the 272H and Tony Thomas from the sales staff was instrumental in its development and introduction.

The designation of the new tractors thus became 262 and 272, the final two numbers indicating the brake horsepower of the engine. The 245 model continued largely unchanged but was fitted with the new internal hydraulic valve chest and a new selector was installed for the PTO operation.

The larger horsepower two-wheel drive, six-

cylinder tractors were fitted with the improved governor, a new draw bar and pick-up hitch and larger tyres – 14-34s fitted to the 285 and 15-34s to the 2100 model.

The four-wheel drive tractors were similarly changed, except for the tyre alterations, and the larger horsepower tractors were all fitted with the new quiet cabs produced in-house. These incorporated sound-deadening equipment with quilting fitted to the side panels and rubber seals at the joints between the cab and tractor tin work, which were to be shortly introduced across the range. This led to the marketing label 'Q' being used for the larger models.

It was claimed that sound transmission was reduced to about 85 to 86 decibels – under the 90

Development of the four-cylinder engine continued and the 262 tractor was introduced in November 1975.

legal limit at the time. The structure and weight of the new cab were considerable, however, requiring sufficient strength to prevent driver injury in the event of a rollover. That meant the former lighter-weight cabs that were rear-hinged to allow easy access for maintenance became a thing of the past. The new cab required two men to disassemble and took up to 40 minutes to remove. The greater sound output of the larger engines was almost certainly a contributory factor in the need for sound-reduction measures

The more powerful 272 was introduced at the same time.

and the relatively insubstantial side panelling of the earlier safety cabs caused vibration and other associated problems. This, together with leakage between the joints between the cab and the tractor tin work, had proved to be a mixed blessing despite the benefits of protection against the elements they offered.

A previously existing problem that was to remain unresolved related to the build-up of internal heat within the cab. Many drivers found it beneficial either to remove cab doors

The 245 continued unchanged because it used the Perkins three-cylinder engine.

partially or to keep them open while the tractor was in use in the summer months. It was to be beyond the Leyland era when air conditioning was optionally fitted to tractor and combine cabs to overcome these deficiencies.

Along with the change to the structure and fitting of the new cabs came alterations to the clutch and brake pedals to make

The equipment needed to remove the cab when safety units were introduced in the mid-1970s. The earlier cabs had been hinged and could be manoeuvred by two men. The new unit required much greater lifting capacity and suggests it would have taken much longer to remove it. Many farms would probably not have been able to afford the lifting equipment and one assumes this sort of facility would only have been available at dealers.

Possible changes to the tractor livery were tried out in the mid-1970s. These are two options for the 272 model.

A further variation.

the driver's access easier. Brakes and the clutch levers were now suspended and used a hydraulic system as opposed to the previous arrangement where mechanical levers passed down to below cab floor level. This change overcame the tendency for sound to be transmitted through the mechanical linkage. Cab pedals still retained the split option of right and left pedals for turning assistance but also had a raised centre pedal to provide full braking, for example for road use.

The steering now became a hydrostatic actuated power unit, previously an option but

Another colour scheme, perhaps less attractive than the others.

now a standard fitting. Again the hydraulic actuation overcame sound transmission – being transmitted through the mechanical linkage, as with the clutch and brake. A new access step to the cab was provided and a floor-mounted throttle pedal provided an option to the hand throttle position. Above the cab, on a separate console, were the controls for the two-speed windscreen wipers, screen blower/demister, interior light and radio control where fitted. All these features were borrowed from the car technology of the time.

The technological changes that had taken place from the 4/65 and 3/45 up until the 262 and 272 models had been to the engine, hydraulics, power take-off, external appearance and cab and rollover safety, but one component had scarcely changed at all: the transmission. With the introduction of the 285 and 2100 models, however, more than double the engine power was being transferred to the wheels than had been dreamt of when the gearbox was originally designed, when it was then transferring little more than 37 hp.

Doctor Henry Merritt, in charge of the tractor design at that time, was a renowned expert on gear technology having worked at David Brown on tractor development and other technical tasks. In the Second

World War he worked on transmissions for tanks, designing what came to be known as the Merritt-Brown tank gearbox. It says something for the suitability of the original tractor gearbox that it was still capable of being mated to much more powerful engines although for a variety of reasons it was becoming a little dated, not least because it still used the original 'crash' design.

The crash-type gearbox made the operation of gear changing much more demanding because it meant that, for the driven gear to be engaged with the transmission driving gear, the speed of the two gears needed to be almost identical to facilitate their sliding together and meshing without the rotating gears 'crashing' into the driven gear at a static or differing speed. Changing gear on a crash gearbox is not unduly difficult when up-shifting but a skill few mastered when down-shifting, often resulting in a horrible crunching noise and damaged gears. That is because the rotational speed of the driven gear is much higher than that of the engine and without synchronising the speed of the two, smooth change becomes impossible.

In automotive terms, until the universal introduction of synchromesh gearboxes in the 1960s, synchronising was achieved through a technique

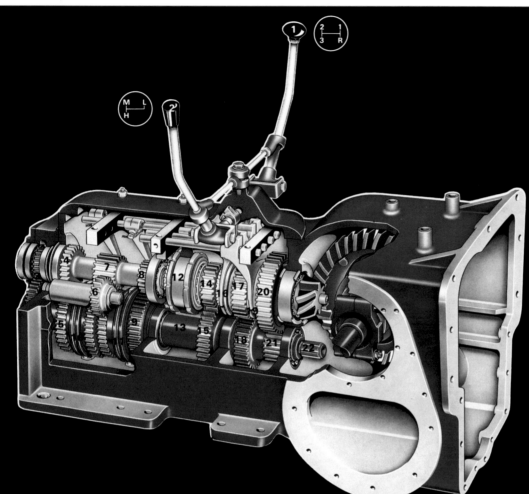

1. Main Gear Lever
2. Range Change Lever
3. P.T.O. Transfer Gear
4. 3rd Gear – Input
5. 3rd Gear – Layshaft
6. Reverse Idler Gear
7. 2nd/Reverse Gear Input
8. 1st Gear – Input
9. 1st Gear – Layshaft
10. 2nd–3rd Gear Synchromesh Pack
11. 1st–Reverse Gear Synchromesh Pack
12. Mainshaft Bearing
13. Layshaft
14. High Range Gear – Mainshaft
15. High Range Gear – Layshaft
16. Medium/High Constant Mesh Coupler
17. Medium Range Gear – Mainshaft
18. Medium Range Gear – Layshaft
19. Low Range Gear Constant Mesh Coupler
20. Low Range Gear – Mainshaft
21. Low Range Gear – Layshaft
22. P.T.O. Primary Shaft

Until the introduction of the Synchro gearbox in 1978, Leyland had experimented with changes to a unit which by then had been in use for nearly 30 years. This sketch shows the layout of the new unit and the way in which the re-working was introduced to produce a nine-speed gearbox with 3 x 3 gear changes. The second view shows the primary and secondary levers for operation of the nine gears and the relationship between the individual gears and the primary selectors shown to the right side of the diagram. This was, of course, the front of the gearbox closest to the drive from the engine.

LEYLAND *SYNCHRO* GEARBOX

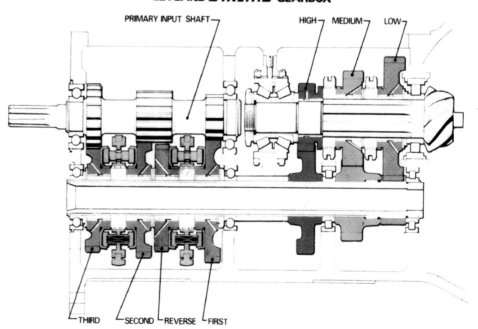

known as double de-clutching. This meant taking the gear out of the driven gear into neutral and 'blipping' the throttle to increase the crankshaft revs. This allowed the crankshaft to increase in speed, which would in turn equate the speed to that of the revolving driven gear and allow meshing to take place. That was not a technique which could be easily undertaken on a tractor, however, but in one sense that was not a great deal of concern.

Much tractor work was carried out in individual gears, which were selected and used for relatively lengthy periods. Thus a farmer going to a field would select the gear most appropriate to the working of the tractor and the haulage of the implement or trailer but when preparing to start work would select a potentially lower gear to cater for the fact that implements were being engaged with the soil or heavy loads were being carted. They rarely needed to change gear on the move. Where there was a problem, of course, was in road work or where tractors were working on hilly terrain and there was a need to change gear more frequently to account for the differences in gradient and load.

The original gearbox had been made more flexible with the introduction of the 10/60 and 10/42 Nuffield tractors in 1964. The models used a separate two-speed gear in front of the gearbox, extending the range from five to ten forward gears. Most contemporary tractors from the 1950s and '60s had gearboxes of similar design, although one or two attempts had been made at producing stepless gearing and pre-selector type boxes which allowed for easier changing when on the move. BMC had experimented with a Lucas torque converter offering stepless changes in the mid-1960s and by the '70s other tractor manufacturers were looking at the use of synchronisers in the gearbox to improve changing on the move. Ford had produced experimental tractors without conventional gearing, although it had then gone back to conventional gearboxes.

The project to redesign the gearbox began after attempts at engaging an outside sub-contractor to produce a new unit failed, principally due to cost. The company had been further hampered by the fact that two of its design team, who might otherwise have had the relevant experience and ability, had left.

At the end of the 1960s, the new tractor design team, which was being assembled at Bathgate, had recruited several young talented graduates. One of these, Simon Evans, was asked by chief designer Ralph Wigginton to assist with the new gearbox.

Constraints were laid down almost immediately because, in order to limit capital expenditure, the existing external gearbox case had to be retained. This was because the company used relatively expensive tooling for machining the faces of the raw castings and similar complex machinery was used for the subsequent drilling and tapping of the various faces of the cast-iron gearbox itself. The gearbox design could not therefore be easily changed without considerable cost, which in itself would increase the potential cost of each unit manufactured. If that were not enough of a constraint, a further challenge was placed on the designers, namely that the gearbox had to be better than any of the competitor manufacturers' designs. A challenge indeed!

The project began in earnest in February 1973. At that time contemporary synchronised gearboxes had put the synchronisers on to the mainshaft, which often resulted in slow and poor shift quality. Evans came up with a radically different design which put the synchronisers on to the layshaft, ostensibly to take advantage of lower inertial loads when shifting which should result in lighter and faster shifts. Initially synchronisers made by the ZF company were considered suitable but, again, cost considerations made their use impractical.

By coincidence, Turners of Wolverhampton was supplying gearboxes for Bathgate-made trucks and at a meeting with its account manager, Peter Guest, which was to follow a meeting with the truck designers, the matter of the new tractor gearbox design was to be broached. Guest attended the meeting with the chief engineer from Turners and the design of the new gearbox was shown. Evans recalls that the chief engineer, after seeing the design, expressed astonishment in a colloquial fashion and left the meeting in a hurry. It transpired that Turners was developing a similar layshaft synchronising system for truck gearboxes and the chief engineer was obviously concerned that his company's ideas had somehow been appropriated from what it regarded as a secret project. Subsequent discussion and examination of earlier Leyland design and details proved that, by an astonishing coincidence, the two companies had independently come up with a similar design. The two firms then co-operated on the project and Leyland was thus able to purchase Turner synchronisers for use in the new gearbox.

The next problem was to get funding for prototype manufacture. This was initially turned down, perhaps because of scepticism due to Evans' relative youth and also probably because, in the difficult financial position the company found itself, a considerable amount of political jockeying for position was taking place for the limited funds that were available for capital projects for the Truck Division. Over a period of time components were stealthily machined and assembled until, by the end of 1974, a simple prototype box had been made. This was sufficient to show that the concept did achieve the design objective of delivering faster and lighter gear shifts, making it the best 'in class' against the competition. Funding was subsequently found by the company for a small range of further synchronised prototype gearboxes.

One of the useful adjuncts of the new design was that the gearbox was able to incorporate a synchroniser on the reverse gear. At that time in automotive practice most gearboxes had no synchronicity on the reverse and, indeed, not on the first gear either. The provision of a synchroniser on the reverse was of great benefit when undertaking loader work, which was by then becoming a more common requirement for farmers. The gearbox was made into a 3 x 3 design with two levers operating the two separate sets of gears. In fact a 3 x 4 arrangement had been considered by Evans but space considerations meant the gear size was compromised and it was thought more practical to retain the existing 3 x 3 arrangement. It is interesting to note that a 3 x 3 arrangement was what had been installed in the Mini tractor almost 15 years before!

On a more practical front, gear lever redesign was

One of the changes for the operator was the location of levers, which had to be placed on the side of the gearbox.

A further view of the additional levers, this one indicating the high, medium or low range of the gearbox.

required to facilitate access in and out of the tractor cab. The existing gearbox lid retained its original design for the reasons described earlier in relation to machining, but changes to the gearbox design were required so two levers could be located one on each side of the seat.

The introduction of the Synchro gearbox was the subject of an award by the Royal Agricultural Society of England and this was made on the first day of the 1979 Royal Show at Stoneleigh in Warwickshire. UK marketing director Ian Wilson, on the right, receives the award from Sir Hector Laing.

To allow on-site testing, a rig was built at Bathgate because previous testing of this sort had taken place at Longbridge. The new rig was therefore of much greater benefit, being located where design was taking place, and it was mobile to make its use more flexible within the factory. The new gear change was tested on this rig and a hydraulic actuator facilitated long-term engagement and disengagement of gears to prove the gearbox's durability.

Evans was able to give thought to another potential transmission issue that related to the planned use by Leyland of a newly designed, Carraro-driven front axle from Italy. Previous four-wheel drive tractors had required the use of separate propshafts to small gearboxes fitted either side of the axle but the new arrangement used a single, centrally-driven propshaft and it was found possible to adapt the new gearbox for the purpose.

Evans came up with an arrangement whereby, in

relocating two of the shiftable gears on to the layshaft, a gear would always be available running at relevant main shaft speed. This gear would power a spur gear to provide drive to the front axle. A drop box incorporating the necessary gears and splined power shaft projecting forward was designed and fitted underneath the gearbox, the spline engaging with the forward driving propshaft.

The tractors fitted with the new Synchro gearbox were introduced to the agricultural machinery trade at a major promotion at Torquay in February 1978 and subsequently to the farming community, from which time tractors would be available for sale, at the Royal Show at Stoneleigh, Warwickshire, in July that year.

The tractor range was again changed in designation with the introduction of the new gearboxes. It henceforth became known as the Leyland Synchro range and consisted of the 245, 262 and 272, which now became known as the 245 Synchro, 262 Synchro and 272 Synchro. Contemporary documents from the company made much of the fact that rivals did not have a gearbox as good. Neither Ford, Massey Ferguson nor John Deere had Synchro boxes, although they had slightly differing forms of transmission. Those that did, such as David Brown, International Harvester in Germany, Deutz and some Fiat models, had synchros but not on all gears.

The Synchro on reverse made shuttle-type operations such as loading much simpler and this would have aided industrial and local authority users, increasing the opportunity of tractor sales to a wider group of prospective purchasers.

A number of additional less publicised changes took place including the use of the stronger 11-tooth final drive pinion, which was now incorporated as a standard feature and not an option. Tinted glass was fitted to the cabs and a

A 272 Synchro tractor in a private collection.

Following the introduction of the Synchro gearbox, tractors carried the name Synchro on the bonnet, as seen on this 272.

The other side of the tractor, again showing the revised designation.

This 272 carries a further livery experiment, with a smaller decal indicating the use of a Synchro gearbox.

In the summer of 1978, two four-wheel drive tractors were introduced. One was the 462, seen at the factory.

new multi-control stalk installed in the cab for the control of traffic indicators, flasher units, headlights and the horn. A number of other minor but useful features to improve driver comfort and ease of operation were also fitted. For the first time, Kleber-manufactured radial tyres could be specified if required.

The benefits of the new Synchro gearbox providing power to a Carraro-driven front axle led to the reintroduction of four-wheel drive and this was applied to the 62 and 72 hp tractors, which became known as the 462 and 472 Synchros. These were otherwise identical in specification to the two-wheel drive models. The complete range of Synchro gearbox tractors, both two- and four-wheel drive, was launched at the Royal Show in 1979. There, Leyland won the coveted Royal Agriculture Society of England Silver medal for its outstanding contribution to agriculture with the launch of the innovative Synchro gearbox.

In 1979, a further change occurred with the

introduction of the turbocharged 98 series engines. This allowed production of a more powerful tractor with an output of 82 hp and this was provided both in the two- and four-wheel drive models, which became the 282 and 482.

During the same year the range was augmented with the availability of a Finnish-manufactured cab made by Sekura named the Explorer. This was in part introduced at the behest of the Finns themselves, who regarded this as a superior cab to that manufactured and supplied by the company. The cab was of a much lighter construction than that produced previously and introduced more glazing with slender components at the corner. This gave the tractor better visibility and an improved shape, and the cab was fitted with rubber mountings to reduce sound and vibration. This gave a more impressive and modern appearance to the tractor compared with those cabs previously manufactured by Vincent and factory-fitted by Leyland.

The two-wheel drive, six-cylinder tractors were also fitted with the Synchro gearbox. This is one of the 285 models.

A view of the hydraulics on one of the larger-capacity tractors. Note the larger size of the draft links compared to those shown earlier.

The other six-cylinder 100 hp model was similarly fitted.

The four-cylinder engine was considerably larger than the three-cylinder type.

A four-wheel drive 62 hp tractor under test with a three-furrow plough.

Tractors for the export market did not necessarily require cabs. A 272 Synchro with oversized tyres is seen at Bathgate.

In 1979, JCB stopped buying skid units and started production of its own in order to satisfy the demand for its range of JCB 3 Backhoe Excavators, which had formed the backbone of its skid unit purchase for many years. JCB had developed its own transmissions and by the end was only buying engines from Leyland. After a damaging strike at Bathgate curtailed engine-making and put JCB's production at risk the firm changed its engine supplier to Perkins.

The following year there was a further significant change with the introduction at the December 1980 Smithfield Show of what was known as the Golden Harvest Range, called by some the

The Leyland-built cab windows could be almost opened fully and this helped reduce the considerable build-up of heat. Notice the lower, sliding, glazed panel which provides access to the control knob that locks the hydraulics and the adjustment to the right-hand link for use when ploughing.

The seat was now not just fully padded but could be adjusted to provide a comfortable position to suit the build and height of the operator.

Automotive rocker switches were extensively used. These indicate various lights and other fittings.

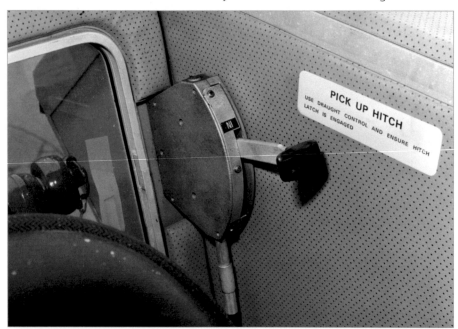

Controls for the pick-up hitch with appropriate health and safety information.

The position of the hydraulic controls.

The brake pedal arrangement. The centre pedal overrides the individual pedal control by providing complete braking. The option of using left or right side braking for more rapid turning during agricultural processes was available by using the lower left and right pedals. Note the flush cab floor.

The improved layout of the binnacle around the steering wheel. The rather large throttle rod appears to be a somewhat less elegant fitment compared with the sophistication of the more modern plastic instruments and surround.

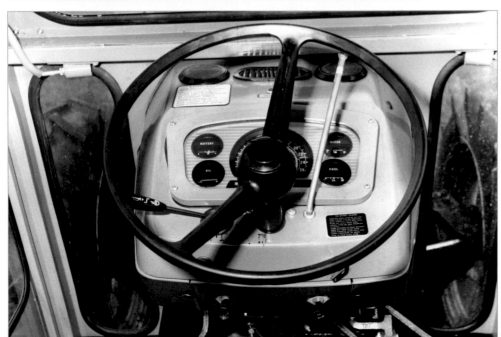

Note the provision of a control stalk on the left indicating horn and basic light functions.

Harvest Gold. This range was chiefly a cosmetic adaptation of the existing tractors but, of course, incorporating all the updated engine and gearbox components which had been the result of significant investment throughout the previous decade. These tractors were now painted in what was an attractive Harvest Gold paint scheme with the new BL logo and repositioned decals along the bonnet.

The ranges were changed to show the horsepower designation of the tractors with the 800 series, actually the 802 or 804 with respectively two- or four-wheel drive, comprising the 82 bhp tractor which had been introduced previously.

The 700 series comprised the 702 or 704 four-wheel drive variant, which replaced the previous 272 and 472. Similarly, in the 600 series the 602 and 604 replaced the previous 262 and 462 models. The smaller tractor, previously the 245, became badged as the 502 and featured the 47 bhp, three-cylinder engine that had been in production for some time.

The hydraulics control had been made more sophisticated with the introduction of the position control several years earlier. This view shows the quadrant with its designated settings for position draft control and the use of auxiliary fitments.

The pick-up hitch fitted to a later-model tractor.

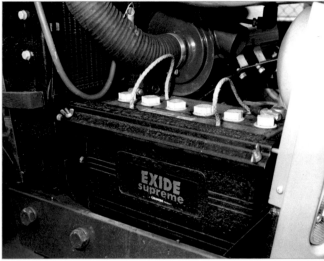

The rear hydraulics with pick-up hitch in position. The lift rods and draft lifts had been enlarged for the use of heavier implements.

The front of the tractor showing the heavy battery, which needed to be located where it could be removed easily. The air filter can be seen above it, together with the bottom of the tank which is obscuring the front of the radiator.

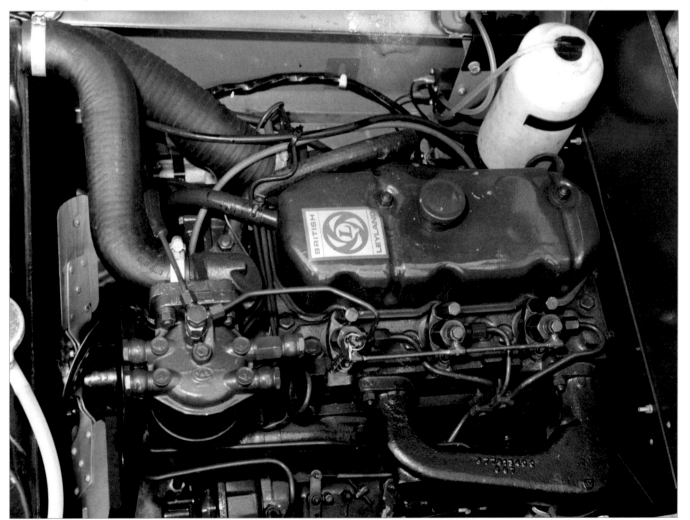

One of the three-cylinder engines, with the top of the radiator on the left.

JCB remained an important customer for the tractor throughout the period of this book. Towards the end of the 1970s, however, JCB was producing its own transmission and associated equipment and consequently it was only the 98 series engines which formed the basis for the last tractors built under the joint arrangement between the company and Leyland. (©BMIHT)

In December 1980, the Harvest Gold tractors introduced predominantly cosmetic changes. A new cab known as the Explorer was sourced from Scandinavia and is seen fitted to an 802 model, which was the new designation for the 82 hp tractor. The smaller 602, with a 62 bhp engine, is fitted with the standard company cab, which had been installed for a number of years.

The Sekura Explorer was a mpre sophisticated cab than previous models. The amount of ventilation through the door was considerable but it did not overcome the fact that the build-up of heat from both internal and external sources was considerable in hot weather. Those wanting more comfort had to await the introduction of air-conditioning which was not produced for tractors until after Leyland had ceased manufacture.

An Explorer cab on a 72 hp 702 tractor.

An 82 bhp 804 four-wheel drive variant at work.

The same tractor ploughing with a four-furrow reversible plough. It is noteworthy that the cab roof is open, presumably to get rid of excess heat.

All the tractors benefited from what were regarded as modern engines, a Synchro-fitted gearbox with oil-immersed brakes, precision depth control hydraulics. Independent PTO with multi-speed options available on the larger tractors. Thus, they fully competed with tractors produced by other manufacturers.

As can be seen in Appendix 2, tractor production dropped dramatically at the start of the 1980s. In 1979, more than 11,250 tractors were built but by the following year fewer than half that number was achieved. There was to be a further reduction in 1981 at the time of this change to the Leyland range.

A smaller two-wheel drive 50 hp tractor using a Leyland-marketed QF loader.

A further revision had taken place to the binnacle and vents were provided either side to the bottom left and right side. Note also the use of a centre mounted multi-function stalk.

A 70 hp Harvest Gold tractor awaiting purchase at an East Anglian auction. Tractors of this sort are sufficiently modern to be capable of being used either in agriculture or as part of a private collection.

CHAPTER 5

Design and Production

The resources available for tractor design prior to the Mini were located at the BMC Ward End factory in north-east Birmingham and a small team had been available to carry out minor modifications as the original Nuffield tractor was developed. These changes included alterations to the frame to suit different engines and alterations to the hydraulic system and linkage to take account of developments and implements. These modifications were modest and only a small team of about six draughtsmen were employed under Bill Watton.

The Mini had seen an outside party, Tractor Research, becoming involved in design development and testing and this had been carried out by a relatively small team. With the completion of Mini work in the mid-1960s, an arrangement was entered into with BMC whereby Tractor Research would undertake further design and development. A range of four tractors was to be produced, of which the Mini was already completed. There was to be a slightly enlarged Mini-sized tractor but in the end nothing came of this because when the Mini had the 1.5 litre B series engine fitted, there was no longer a requirement for it.

The medium and large tractors, however, were a different proposition. In the mid-1960s the 10/42 and 10/60 were the final development of the original design introduced in 1948 and were starting to look dated. Something different was needed and this resulted in more extensive involvement by Tractor Research. A number of its employees were relocated from Coventry to the design office in Longbridge, where they became part of a larger team of designers producing new vehicles for the entire BMC Group.

Although the 3/45 and 4/65 tractors were not considered to be particularly successful, BMC tractor staff thought Tractor Research did not have enough time to fully design out various problems as they occurred. Almost as soon as the tractor had been introduced a raft of technical changes and improvements took place to overcome these deficiencies.

By the time these changes had been worked through, the merger of BMC and Leyland had taken place and decisions being made by Sir Donald Stokes extended as far as tractor production, design, sales and marketing. One of his edicts was for the designers to alter the tractors' appearance, the most obvious of which was the change of colour to the successive range of 384, 344 and the 4/25 Mini, which became the 154, and the badging being switched from Nuffield to Leyland. A number of further technical changes were introduced to the medium and large tractors, which have been shown in Chapter 4.

Sir Donald's involvement was much more radical in that he wanted to set up an in-house design team and this was to be based with other ancillary tractor functions at Bathgate, where, at that stage, tractors were only being manufactured. As with other parts of the organisation, such as sales and marketing, new blood was to be introduced and Ralph Wigginton was recruited from Standard–Triumph.

Bill Watton was placed under Wigginton for a short period but the decision to relocate the design office to Scotland almost certainly did not go down well with him and he did not move to Bathgate.

The senior designer under Wigginton was Mike Barnes who, after a career with a hydraulics company and then David Brown, joined Leyland in Birmingham in December 1968. At that time he worked in the Longbridge design office, which was shortly to transfer to Scotland although not without some dissent from staff. Wigginton was to remain in charge of design until 1978, when Barnes took over.

A small number of graduates were also added to the design team. Two of these were Welshmen: Simon Evans, who went on to design the Synchro gearbox, and Bruce Davies, who was involved in hydraulics development. In addition, Tony Moore joined and became extensively involved in testing. Further recruits included Terry Stephenson, from International Harvester, and Phil Young, from David Brown. Tom Queen, who was a former British Railways carriage engineer who then worked for the Rootes Group, became involved in early cab design as did Davies, who was extensively involved in developing prototypes for the Q-cabs. The original 'basic' cab provided protection against rollover but the subsequent Q-cab introduced in 1976 had significant sound reduction to comply with newly-introduced noise regulations.

One of Barnes' responsibilities was to attend a monthly meeting held in Preston by Leyland's Product

Construction of the BMC bus, truck and tractor plant at Bathgate. The nearest building became C Block where tractors were built. (©BMIHT)

Improvement Committee. Tractor business was generally conducted at the end of a day of meetings attended by truck and commercial vehicle teams and all too often there was little time left for it. This led to frustration at the lack of interest and consequent low level of financial investment in the tractor business. Ideas were forwarded from the field sales team, who learned at first-hand what the consumer – the farming community – wanted and channelled feedback through marketing director Bob Turner.

The feeling grew that the tractor activities were very much a lesser part of the merged company. This sentiment was no doubt exacerbated by the financial difficulties of Leyland at the time and eventually the knowledge that a potential sale of the business might be considered.

The design office was initially located in B Block at Bathgate but subsequently moved to a former garage on the other side of the main Edinburgh to Glasgow trunk road. The garage's front showroom was converted into the design office and development took

place in the workshop facilities. These facilities were to be expanded and also used by truck designers.

The Bathgate factory had been constructed by BMC with government funding. In the 1950s, the government was concerned there were large pockets of relatively high unemployment which would be difficult to eradicate without direct intervention at a national level. Consequently, it decided that large-scale manufacturing should be relocated to areas of high unemployment. The construction of the Bathgate factory fulfilled that requirement because it was in an area of significant joblessness caused by the closure of a large number of coal mines in the vicinity.

BMC acquired a 266-acre site on the outskirts of Bathgate and in 1959 construction of the factory, designed by a Birmingham firm of architects, was started. Construction was finished a couple of years later and tractor production began on a relatively modest basis towards the end of 1961. Some component manufacturing remained elsewhere, with the three-cylinder and Mini engines still

Part of the truck assembly plant at Bathgate, showing its large size. (©BMIHT)

manufactured in the West Midlands. Other component and major castings continued to be provided by firms both within and outside the group. The company foundry at Wellingborough made large transmission housing castings. Some tractor activity initially remained in other parts of the country, principally sales in Oxford and design in Birmingham, but these functions were later moved to Scotland.

The factory comprised four separate blocks lettered A, B, C and D, of which C was used for tractor production. Block B was the location for engines and large-scale machining for tractors, trucks and buses, which were made in Block A. Block D was devoted to disassembling completed vehicles for export and in the case of tractors this meant stripping down various parts and packing them into crates so that reassembly could be carried out in the destination country.

Components such as wheels, tyres, fuel pumps, instruments, hydraulic components and sundry items continued to be purchased from outside suppliers. The cabs were initially manufactured by Vincent, those for the Leyland 154 coming from Winsam, a large-scale manufacturer of cabs for many years. Sheet metal for such items as the bonnet and mudguards was produced by a firm at Willenhall in the West Midlands.

Engines and components were tested on site, each engine being subjected to a 40-minute procedure during which a partial load test was applied for a minimum of 20 minutes, with a further 10 minutes provided at full load. Heenan and Froude dynamometers were used for engine testing as had been the case when engines were manufactured in Birmingham.

A four-square rig was produced for gearbox testing and was used when the Synchro gearbox was manufactured. Tractors were subject to more exhaustive testing when new components or other

Components were resourced from throughout the company's in-house suppliers, among the largest of which was the Wellingborough Foundry. This produced parts such as large gearbox transmission castings. (©BMIHT)

developments were added and this was usually carried out on local farms because it was considered that working under such conditions was more realistic than using rigs within the factory. The hydraulic lift was subjected to separate testing to ensure performance specifications could be met.

When tractor production started in the early 1960s, assembly was carried out on stands in fixed positions. When the workforce was sufficiently trained, a moving line was installed and tractors were built up on a continuously moving track. A spray booth had been installed for painting machines when they were part-assembled. Components were painted in the lighter blue or silver tinwork and the wheels were fitted afterwards.

Tractor production had originally been set up for 1,000 machines a week but for much of the production period an average of about eighty tractors a day, or 400 a week, was achieved. This indicated a total of about 20,000 a year but, as can be seen elsewhere, rather fewer than this were produced within a few years.

At one stage it was considered that about 50,000 a year would be made to satisfy the demand for medium and large tractors, with a considerable proportion of the 1,000-a-week production taken up by the Mini. BMC had undertaken market research into potential Mini sales and considered that about 400 tractors a week would be needed to satisfy demand. However, poor sales meant this total was never achieved. In fact, the largest number of Minis assembled in one week was 135 and the least was two. Initial production was between twenty and forty per week and it was understandable that numbers would inevitably be small until the Bathgate workforce was fully trained. Production continued in Birmingham until it was taken fully over by Bathgate, although this was before the Leyland merger.

An early Mini on the assembly line. In the early days of production, tractors were built up on semi-fixed stands and individual machines were made as required, leading to different models being built adjacent to one another.

Leyland changed its manufacturing methods when assembly skills improved and this led to the introduction of a rolling line. This tractor for Finland was one of two painted silver.

The best day of production amounted to a hundred, of which ninety-eight were passed, including skid units. A more typical daily total was eighty but that was not always achieved. About 100 personnel were involved in assembly. Tractors were dispatched from the factory on trucks but a rail siding was connected to the national network on which tractors could be put on to flatbed wagons and taken to locations from where they could be exported. That of course did not apply to those which were crated in 'knocked down' form.

When technical changes were introduced the testing could be extensive and a facility was provided whereby a tractor could be connected to a radial arm fixed to a central anchor point and run driverless around a circular track. This would allow continuous testing of the tractor and these tests could go on for several days

Bathgate's D Block was given over to packing tractors for export. Note the 154 fitted to its timber stillage and ready to be wrapped. Other larger tractors were packed to make the maximum use of containerised shipping and distribution.

so that a reasonable amount of hours and potential problems could be identified in a relatively short time. Despite these tests, small but irritating faults could be found, such as the incorrect fitting of the control quadrant leading to complaints that the hydraulics were not working correctly on a newly delivered tractor. This required a visit by a member of the sales team to carry out testing with assistance from the dealer's technical staff to find and cure the fault. Other manufacturing faults could be more significant and a Sussex agricultural contractor recalls that a batch of tractors were dispatched with the transmission and engine power shaft poorly aligned, leading to significant repair being required shortly after the tractors were put into use.

Sales staff could be involved in 'unofficial' improvements. When the hydraulics were found wanting in lifting some of the heavier implements being introduced, a simple design of assister ram was developed by Brian Webb and a dealer engineer and made

available to provide extra lift capacity. This was before the company began offering such rams as optional extras on factory-built tractors.

Finland was to become one of the most important external markets and the requirement for substantial insulation against the weather, apart from other considerations, hastened the involvement of Sekura, a Scandinavian company, in the design of the cab. The

The end of the assembly line, reputedly showing a batch of 4/25 tractors. These were produced for Mexico and representatives from the purchasing authority are seen examining a completed machine. The member of staff third from the left is Bob Turner, who was in charge of marketing, and his presence presumably indicates this was a fairly important customer.

Finnish market also required more powerful tractors than were being produced by the company and the six-cylinder engine was developed at least in part for this market, but also for Australia where the larger farms dictated the use of a higher horsepower tractor with more substantial implements. Finnish tractors were frequently used in forestry and similar work.

Development of the factory continued through the 1970s with additions to a number of the units.

Consideration was given to a new assembly building to make cabs and sheet metal components for both trucks and tractors. At the rear of the Mosside training school an area of ground was levelled and foundations dug for this new building but it was not built before the ending of tractor manufacture in 1982.

In the 1970s, concerns were being expressed regarding vehicle emissions and the need for more fuel-efficient engines. The company had entered into discussions with the North American engine manufacturer Cummins and some consideration was given to erecting a separate engine assembly plant at Bathgate. In fact, Cummins engines were eventually produced in Livingstone in Scotland and went on to be extensively used, not just in agriculture but in other commercial vehicles. The Leyland tractor was never fitted with such engines in production.

The assembly plant in Turkey at an early stage of its development in the 1960s. It was a venture set up jointly between BMC and the local distributors, of which there were two principal but separate organisations in that country. The factory also produced commercial vehicles.

BMC Turkey

In the post-war era, Turkey had for some time been one of the best export markets for BMC. There were two principal distributorships which separately imported vehicles under the Austin and Morris name. When tractors were exported they were painted in different colours to reflect this.

In 1966, with assistance from BMC in the UK, a Turkish plant was constructed which assembled imported vehicles of all sorts in mainly Completely Knocked Down (CKD) form. Again, the principle of resourcing this with locally produced items was continued. Eventually tractors were introduced as part of this arrangement and a greater proportion of completed vehicles were made from local resources.

The B series engine, which was used in the 4/25 and subsequent Leyland 154 model, was assembled in Turkey and tooling was

The inside of the Turkish plant, showing castings being produced on a continuous assembly line and the cupolas being tipped to discharge hot metal into the moulds.

subsequently exported to the country so the complete engine could be made. An enlarged version of the engine with 1.8 litre capacity and 30 hp output was produced and this was available following discontinuation of the 154 in 1979. The tooling to build the actual tractor went to India, including the 1.5 litre engine. A number of these engines were re-imported into the country but were assembled from parts manufactured overseas because of concerns about the quality of fully assembled products. Most of these engines were sold into the marine trade, where they were used in narrow boats and other similarly sized inshore vessels.

The final Turkish-built model produced for British Leyland was the 302, which became the smallest tractor in the Harvest Gold range.

Towards the end of the 1970s it was becoming increasingly apparent to management in the tractor organisation that, with the dire economic situation not just inside the company but also in the national economy, there was little likelihood of there being a long-term future for the tractor. Funding was found to develop a more cosmetically attractive range and the result was the Harvest Gold series. These tractors incorporated the recently developed Synchro gearbox and the successful 98 series of engines, which had been developed further from the initial units produced in 1972.

It seems somewhat tragic that, although the company now had a product that was equally as good, if not better, than that produced by its competitors, this situation had only come about because it was likely the tractor business was to be sold or merged. These changes were being put in place for the purpose of achieving the best price for the business, rather than for continuing it under the Leyland banner.

There is little evidence as to the nature of the discussions that took place regarding the disposal of the business. There had been interest by some company employees in a possible management buyout but in the event it was sold to a wealthy Lincolnshire businessman, Charles Nickerson. He had previously bought Track Marshall, another British Leyland subsidiary, and therefore had a record of commercial dealings in tractor production and sales, albeit on a limited scale.

Nickerson had previously purchased the former Britannia Iron Works at Gainsborough in Lincolnshire and tractor production was transferred there, although outside the ownership of British Leyland and with a range of products which was to become very different from that which had been produced by the company.

CHAPTER 6

Development and Testing

Tractor Research, the designers of the Mini tractor, had a strong background in development testing because that was one of the primary activities of the parent company, Harry Ferguson Research. The Mini, being largely an entirely new tractor with several innovative features, took several years to develop fully and bring to a state where it could be manufactured.

The components and complete tractors were tested in-house at Tractor Research's headquarters at Toll Bar End, Coventry. The test operations were filmed and incorporated into a promotional film called *Seeing is Believing*, which BMC released at the time of the launch. Components are shown being subjected to vigorous trials, including inclinometer tests for both the engine and transmission to ensure that oil starvation did not cause premature failure. Both vertical and lateral angles could be adjusted on the rigs so a reasonably wide variety of tests could be applied to these components, which they passed. Endurance testing of the hydraulics was also undertaken on a special rig.

Field tests were carried out on a farm at an estate near Charingworth in the Cotswolds and some contemporary amateur film has survived of these showing a prototype at work. This machine was tested against other tractors acquired by Tractor Research, including a Porsche Junior and a French-built Massey Ferguson 825. In addition, three President/Stokold tractors were used for comparison purposes and, as has been explained previously, the hydraulic system and three-point linkage was developed on one of them.

These tests were the forerunner of a more substantial group of trials subsequently carried out with direct involvement from BMC prior to the tractor launch in 1965. A farm near Knowle in Warwickshire was used for specialist tests in very wet conditions because it was anticipated the tractor would be used in export markets for work where such conditions were normal, such as paddy fields.

Further tests of a more general nature took place at a farm at Enstone in the Cotswolds. This was behind a restaurant known as The Quiet Woman, now an antiques centre of the same name. Here, a further range of extensive implement tests were carried out and at least one of the engineers seconded to this

One of the second batch of prototype Minis, now with the correct tin work for the production tractor, being fitted to a radius arm to allow continuous operation. Note the concrete obstructions, over which the tractor was to be tested.

An important test from the point of view of marketing the tractor was that held under the auspices of the NFU in Scotland before the launch of the 384 and 344 models. This is the start of a period of round-the-clock ploughing which involved the cultivation of a significant area of farmland.

work was Jeff Kitts, a BMC employee. Further tractors were tested at the firm's Toll Bar End workshops and a pre-production model was fitted to a merry-go-round, where it was subjected to lengthy procedures. These pre-production prototypes appear identical to those subsequently manufactured for sale and had features that differed from the initial prototypes.

Several years after developing the Mini, Tractor Research set up a small development facility at a farm at Charingworth. This was more convenient than having to return tractors continually to Toll Bar End for modification and maintenance. The main Harry Ferguson Research business subsequently moved to the former Armstrong Siddeley car factory near Coventry, but this occurred after Tractor Research's involvement in tractor design and development for Leyland.

Less is known about the development of the 3/45 and 4/65 models tractors but similar field tests took place as those undertaken near Charingworth on the Mini. A contractor tractor driver, Richard Cole, who now restores old Fordson tractors as Cotswold Vintage Tractors, tested the Mini on a number of occasions. He was told to be at Charingworth on one occasion in late 1968 because something interesting would be happening. After his arrival in the morning, a sheeted

tractor arrived on the back of a British Road Services lorry. When it was uncovered, a Nuffield tractor was revealed, probably a 4/65, which was fitted with a blue-painted V8 engine, probably a Perkins unit. This was put to test ploughing with astonishing results. Cole reported the tractor was driven at some speed along the furrow with soil literally flying over the tops of the mouldboards of a six-furrow Ransomes semi-mounted plough. He later ploughed with the same tractor with similar results. The engine delivered considerably more power than the clutch could transmit to the gearbox, with the consequence the tractor was frequently stopped for clutch changes.

Tractor Research continued development work for BMC but after the Leyland merger the arrangement came to an end after design and development work on the Leyland 344 and 384 models. After 1969, the design and development functions of the combined company were moved to Scotland.

The new 344 and 384 models were subjected to an extensive endurance test which one suspects the relatively new marketing manager, Bob Turner, had no small part in proposing for publicity benefits. This pre-production field test was conducted at Ratho, west of Edinburgh. Turner had probably previously been involved with, or certainly knew of, a similar range of

A test tractor surrounded by various Leyland personnel, including Vincent de las Cassas, who is leaning on the plough and apparently talking to Bob Turner, who was in charge of marketing and general tractor operations.

tests carried out when the Fordson Dexta was introduced in 1957. As an aside, the Dexta was first exhibited to the agricultural trade at Alexandra Palace in north London where the 384 and 344 were launched, albeit 12 years later, one suspects with the encouragement of Turner.

The trial was held under the supervision of the National Farmers' Union of Scotland and two tractors were used, supposedly the first two off the production line. After some bedding-in of components and other minor adjustments the machines embarked on an exhaustive ten-day trial, working continuously day and night. The prolonged test needed considerable organisation by Leyland and required that tractors changed crews while on the move. Refuelling also took place in a similar manner, fuel being provided from a bowser hauled by a remodelled Leyland 154. The same tractor was used for marking out the headlands for the test. The trial was considered a

success, with more than 800 acres ploughed, and the company heavily publicised the event.

Technically, however, the long periods of continuous use caused significant problems. Issues with the hydraulic systems led to tractors being taken out of use and quite extensive repairs were required to keep them operational. The hydraulics seriously overheated and transmission oil was found to be boiling. When the rear transmission casings were opened to examine the internal components this caused oil to splash on to the technicians' boots. In the later stages of the trial this attention became a nightly requirement. The cab interior was also found to become unbearably hot, requiring the doors to be kept open despite the chilly weather.

The repairs had to be carried out with some secrecy given the importance of the trial. Indeed, a promotional brochure that followed the tests made much of the fact that ploughing took place not just

The personnel involved in the continuous test. From the worn appearance of the overalls and the smiles on their faces one assumes they were pleased to have completed their task.

throughout the day but also at night and involved a large team.

In the late 1960s, Tony Moore was one of a number of graduates taken on by Leyland and located to Bathgate where the majority of ancillary tractor activities were to be based.

Moore has described in some detail his activities with the company and it is interesting to learn what steps were taken to ensure that particular models were developed fully prior to being put on the market. One of his first tasks was to test the new 253 tractor introduced in 1971. This was fitted with a three-cylinder Perkins-built engine because there was no longer a three-cylinder company-built engine available as the tooling to produce it had not been relocated to Bathgate.

The National Institute of Agricultural Engineering had a long history of independent testing of tractors, although with the passage of time this became less significant as individual manufacturers took over a number of the operations. One of the last Leyland models to be tested was the 270, seen with oversized tyres presumably at the institute's facility at Silsoe near Bedford.

O.E.C.D. Approval 473
Report 638
November 1974

REPORT ON TEST IN ACCORDANCE WITH O.E.C.D. TEST CODE FOR AGRICULTURAL TRACTORS

LEYLAND MODEL 270 TRACTOR

Test No: R.74/7786/O.E.C.D.

Manufacturer: British Leyland (U.K.) Limited,
Bathgate,
West Lothian, Scotland.

NATIONAL INSTITUTE OF AGRICULTURAL ENGINEERING
Wrest Park   ·   Silsoe   ·   Bedford

Moore was involved in the testing and development of the first prototype, which was running by February 1971 and was to be followed by three further pre-production units. The prototype was used for the initial performance tests, including those on the hydraulic system, but for endurance field testing a contractor was used. Later, Moore took over some of the endurance work and also undertook performance tests. By May that year, cooling performance tests had been completed in the prototype test cell in the factory and Perkins had given its approval to the development thus far. The tractor continued to run reliably and on 5 August it was agreed to launch the 253 at the Smithfield Show the following December. The three pre-builds were assembled in C Block and these were to go on display at the show. Following their use for publicity purposes, these tractors were to be used for additional testing and Moore was involved in carrying out development work for two makes of loader. In addition, he carried out traction trials with a range of differing tyres.

The Bathgate factory had an external area where tractors could be fixed to a radius arm. This was connected to a centre pivot set into a substantial concrete foundation and allowed the tractor to be set to work under its own power and rotated without the use of a driver for long-term endurance testing. This could be used for a variety of applications and the tractor could be driven over obstructions with implements attached. A typical test could last for 500 revolutions to simulate testing under arduous conditions, which could equate to such use over the life of the tractor.

This facility was augmented by a hydraulic test rig area, which was adjacent to the experimental workshop

A prototype six-cylinder tractor carrying out a cultivation test at a farm in Scotland.

A further view of a similar test. Note the substantial semi-mounted plough being used. Rigs were available in the factory for component testing.

A prototype 2100 100 hp tractor being tested with a rotavator. Lengthy trials with equipment of this sort enabled the tractors to prove their suitability quickly.

in B Block where other development work had been undertaken. It was here the position control system for the hydraulic power lift was tested after many hours of field trials to try out different sizes of hydraulic valve. At the edge of this area was a test chamber in which air temperature could be raised to simulate tropical conditions and within this facility there was a PTO dynamometer for performance and endurance testing.

Thought was being given to data collection in the field and a system was devised using a portable ultraviolet recorder housed in a cab extension, which could be bolted to a test tractor. The system worked well and was regularly used to measure structural loading (strain gauge data), dynamic hydraulic pressures and response of tractor systems.

At Mosside Farm, a four-square rig for transmission tests was installed. The rig is described on page 80 in conjunction with the testing of the Synchro gearbox. This structure consisted of two tractor transmissions placed back-to-back and, instead of wheels being fitted to the axle flanges, large-diameter chain sprockets were installed. High capacity chains could be used to join the outputs of the test transmission and the slave transmission to determine the suitability of the gearbox under test. Different tests could be carried out with selected gears in the development gearbox being used to drive the same gear, albeit in reverse in what was the 'slave' transmission, i.e. that being driven by the transmission under test.

Moore was subsequently involved in safety and quiet cab development at Bathgate. A dynamometer car was used to load the tractor so that tests could determine the noise output and the suitability of the cabs in reducing it. This was a major exercise under the supervision of Bruce Davies, another of the late 1960s graduate entrants. Numerous concepts were designed and tested to ensure compliance with the, by

A 4100 during a forage harvesting trial.

Testing the hydraulics of a prototype tractor. The bearded engineer is Bruce Davies, who was one of a small band of graduates who joined Leyland in the late 1960s and who were extensively involved in a number of functions including design and development throughout the following decade.

then, stringent legislation requiring cabs not just to provide safety in the event of a rollover but to reduce increasing noise from the mechanical systems. The early tractor cabs from the 1950s and '60s were originally designed to provide a degree of weather protection but this was often at the expense of noise and vibration and provided little in the way of driver protection in the event of overturning.

Moore was further involved when radial tyres were being introduced in the late 1970s, as described in

Chapter 4. Pirelli had made considerable claims for the benefits of the tyres but Leyland was keen to know if these could be substantiated. Moore was involved in testing using a six-furrow, semi-mounted plough to which he fitted a load cell into the headstock. This would permit measurement of the pulling loads. In addition, a torsional shear box was developed to measure the strength of the soil. The plough permitted raising or lowering of differing bodies into the soil to permit its use behind different sizes of tractors. When the tests were completed it was found the radials showed approximately ten per cent less slip than the cross-ply tyres that had formerly been in use.

Leyland decided it wanted to submit a tractor, a 270 model, for the Organisation for Economic Co-operation and Development (OECD) tractor performance test. Prior to this, considerable factory testing took place to ensure the tractor concerned would perform to its maximum ability.

Moore was involved in development of the 2100 model, a two-wheel drive, 100 hp tractor, and an initial paper evaluation took place for a suitable transmission to go with it. However, Leyland, as had so often proved to be the case, was very short of capital for developing an expensive component such as a new gearbox. It decided on a low-cost solution which involved minimal development of the existing unit and only three prototype tractors were to be made.

Moore spent time using one of these prototypes for ploughing and in addition he obtained a 100-inch Howard rotavator to test it fully. Publicity shots were taken at Mosside Farm and testing was carried out on farms in the area. Later, Moore was involved in further plough tests at Riccarton Farm in the Bathgate area.

Testing could be undertaken on a less formal basis. After the introduction of the tractors with the Synchro box, Tony Thomas, who was then the product training and demonstration manager, took the opportunity to drive one of the new Synchro-fitted 272 tractors within the factory and later that day at Mosside Farm. Thomas was very impressed with the smooth gearbox,

more so than with any competitor he had driven, but he felt there were some inadequacies with some of the ratios available for farm work. Bob Turner told Thomas to go to Perthshire where a prototype was at work under the supervision of the local distributor, Marrs of Burrelton. He was accompanied by Brian Webb, who had worked with Thomas on similar testing of the pre-Synchro ten-speed box which had led to the introduction of the H model 262 and 272.

The pair concluded that, with the exception of bottom and top gear, the intermediate ratios were not suited for agricultural work, be it arable or grassland. What exacerbated the unsuitability was the continued use of the nine-tooth final drive pinion, which had a high rate of failure particularly in arable use, both at home and abroad. A critical report was submitted to Turner, and Thomas and Webb were asked to recommend suitable ratios for the gearbox based on the use of the eleven-tooth final drive. These were subsequently agreed to by Ralph Wigginton, who was in charge of engineering, and production tractors were sold with the revised ratios. Both Thomas and Webb had previous careers in practical farming before joining BMC as demonstrators in the early 1960s and thus were ideally suited to this task.

By the late 1970s, the tractor business was beginning to be affected seriously by the company's dire financial state and significant technical development was no longer an option. The introduction of the Harvest Gold range was more of a cosmetic improvement incorporating the new Synchro gearbox and an improved 98 engine, but benefited from the use of oil-immersed brakes, a change which had been sought for a while. Information on testing is sketchy for this end period and one assumes that field testing of the sort described previously would have continued.

With the company considering the option of disposing of the tractor business there was very little money available for improvements other than minor changes thought likely to help obtain a better price for the business.

CHAPTER 7

Sales and After-Sales

Sales, service, after-sales, warranty and training activities remained remote from the Bathgate plant after manufacturing was transferred to Scotland. In the early 1960s, the factory had been under the direction of Keith Sinnott, who was managing director of BMC Scotland. Sinnott had a long career in tractor sales before his Scottish appointment but was to hand

over the operation to Raymond Smart. Subsequently, Sinnott was appointed director of home sales and export at Longbridge in 1966, a role he carried out for three years until becoming director of tractor sales in 1969.

Sinnott's previous assistant, Tom Cummins, was chief sales manager and based at Cowley, where sales were located in the early 1960s until BMC transferred this function to Longbridge. Under him was a small team in charge of different regions of the country. The company ethos at BMC at that time was one where older traditions of business conduct held sway. This ethos was bolstered by the fact that at that time sales were often four to five months ahead of manufacture and, consequently, there was no great need to pursue orders more vigorously because the factories could not cope with the extra production.

The sales function was relocated to Longbridge in the mid-1960s and was to stay there until 1969, a year after BMC and Leyland merged.

When Sir Donald Stokes started to reorganise the company in 1968 he identified that tractor activities such as sales, after-sales, design, development and training were

organised rather disparately throughout the West Midlands and to the South. He intended to rationalise these arrangements by transferring as many of these activities to Bathgate as possible. Sir Donald further determined to bring in an outsider to manage marketing and associated activities because he saw the need to ensure the sales organisation became a rather

During this period the use of dealers' labels and plates identified the supplier, and, in this unusual case, what appears to be the serial number. Collings Brothers sold considerable numbers of Leyland tractors through the years.

Not all labels were quite as elaborate and throughout this period less expensive plastic stickers were used, a somewhat less attractive feature than the cast plates of previous years.

more active regime than he thought had been the case in the BMC era. He recruited a former Ford employee, Bob Turner, who, for the next ten years was effectively in charge of marketing and sales, although he also spent some time in similar truck-related activities. Turner was to have a significant say in the direction of the tractor business.

Turner was a senior figure in Ford's tractor division during the 1950s and '60s. He had started his agricultural career at the Henry Ford Institute of Agricultural Engineering at Boreham near Chelmsford in Essex in 1948, where he took a general course in engineering, animal husbandry and farming practice. In the early 1950s, he joined the Ford tractor division at Dagenham under managing director Bill Batty. He rose quickly through the ranks and became a sales representative, visiting dealers in the south of England, from Kent to Wiltshire.

Bob Turner at a meeting with a dealer, Sparshatts of Winchester, where a new sales agreement appears just about to be signed.

Not all tractors were reduced to basic components for export. This tractor appears to have been partially stripped down and loaded on a flat railway wagon for transport to the docks.

He subsequently became manager of the Ford Mechanised Farming Centre at Boreham and had a strong influence on product training policy.

In 1962, Turner was appointed general sales manager for the UK under Geoff Buckley, executive director of Ford Tractor Operations. This was a key period in the company's history when the Fordson Major and Dexta were due to be replaced by the Ford 1000 series that was to be built at a new plant in Basildon in Essex. Turner supervised the closing down of the old Dagenham line in the mid-1960s and the launch of the 1000 series tractors later that year. Ford entered 1965 at what was a difficult time for its tractor division because product reliability had become an issue in the early days of the new models.

In 1968, Turner left Ford to join BMC's tractor business as general sales and marketing director, a decision that was something of a surprise to colleagues at Ford. He remained with the group until his retirement and he died in December 2007.

Turner's sales and marketing approach was very different from that which had held sway at BMC until the late 1960s. He had come from a company where marketing and sales were pursued much more aggressively and at the merged British Leyland there was now a requirement to ensure that maximum output was achieved, not just because of the need to improve profitability but also to help arrest the serious financial difficulty which the company faced. By now Leyland was being backed financially by the taxpayer and there was an obvious need to ensure it proved itself worthy of such support.

The effect of Turner's pro-active management style was to put much greater pressure on those under him, something of a baptism of fire for some of the staff. Cummins continued in charge of home sales following Turner's arrival. He died in March 1969.

Tractor sales faced additional difficulties because by the late 1960s it was obvious that the Mini tractor, for which the company had great hopes, was a commercial failure even when fitted with the more

Some promotional equipment was exotic to say the least. This Mini was mounted on a revolving display at Longbridge. (©BMIHT)

powerful engine. A further problem arose in that the medium- and large-sized tractors 3/45 and 4/65 were not selling well, largely because of reliability and technical problems. There was also a general dislike among the farming community of the changed external styling, which even others in the tractor business thought unappealing.

In 1969, Sir Donald's changes saw the sales organisation moved from Longbridge to Bathgate and an office was set up in a group of portable buildings at the end of C Block, where tractor production was located. In the 1970s sales moved to Wester Hailes in south-west Edinburgh before a further move was made to the Guild Centre in Preston, where the department was to remain in a building occupied solely by British Leyland until the tractor business was sold in 1982. Thus, within the space of little more than 12 years, the sales organisation had been based in four locations.

Overseas events such as this demonstration of the Mini in the Far East could lack a little of the sophistication of those in the UK.

Following Cummins' death, Peter Warren was brought in as senior sales manager but he stayed for just a short time before being followed by Ian Gibson in 1971. Gibson was replaced by James Taylor, a former Ford export manager, but he did not stay long either. His role was taken over by Steve Herrick from truck marketing after Turner's departure in 1979 and

he remained in position until 1982 when the business was sold.

If the changes of personality and sales department locations appear to be rather more numerous than one might expect, then the difficult financial situation in which the company found itself, which was continuing to deteriorate throughout the 1970s, was

Leyland was keen to involve itself in the local community, as shown by this line of tractors and equipment processing down Princes Street in Edinburgh.

One of the primary sales venues for tractor manufacturers was the Smithfield Show, which usually took place in December. The BMC stand in 1966 has the new Mini on display. (©BMIHT)

probably a factor. From the employees' point of view, the difficulties the company faced would have been a factor in concluding that a career outside British Leyland was probably more secure than one within it.

Underneath the senior sales staff was a national sales network which was organised into a number of regions. In the 1970s, the south was run by Rob Runciman at a plant in Aldenham in Hertfordshire where, by coincidence, the Routemaster bus had been constructed. Tony Thomas covered the north of England and Scotland from Bathgate.

Brian Webb, who had joined the company as a demonstrator in 1963, worked on training and service-related technical advice until he was put in charge of the East Anglian region and based at Stratford-upon-Avon until 1979. Webb then took charge of West Midlands sales, which was based at Oldbury, west of Birmingham, in the UK Service division. There he worked under Keir Wyatt, who had a national responsibility for service-related activities for a number of years.

The introduction of new models usually took place at the Royal Show at Stoneleigh in Warwickshire, the Smithfield Show in London and at the Royal Highland Show. The stands, tractors and other facilities were provided by the company, although sometimes machines would be borrowed from local dealers. Tractors needed to be presented in as good a condition as possible, so particular care was made to ensure the paintwork and external appearance were at their best.

Throughout the country, agricultural shows had been a feature of local life for many years and the company was represented at them by distributors who provided staff, tractors and other facilities from their own resources. The company had a catalogue of promotional gifts and other such paraphernalia which could be ordered to assist in this. In such situations tractor sales would be pursued actively, although at a national level the three main shows were not normally regarded as ones where direct selling took place. Indeed, the company's presence at national shows was in many ways regarded as defensive in the sense that, if a company failed to take a stand, competitors would

Tractors could be used for more whimsical purposes. The tractor driver at this event at either Thruxton or Silverstone is Leslie Crowther, a well-known television compère and comedian in the 1970s and '80s.

A celebrity line-up at the same event. It is thought the driver nearest the photographer is the singer Engelbert Humperdinck.

Events where tractors used implements were probably more useful in making sales. This 345 with a three-furrow reversible plough is giving a demonstration in north Warwickshire. (©BMIHT)

Leyland produced what was called a Jubilee Tractor in 1977 and painted it a rather attractive silver. It is seen on display at an agricultural show

A more conventional form of promotion was the use of tractors that had been sectioned so the inner workings could be seen. This 344 is probably on display at the Royal Show in Warwickshire at the end of the 1960s.

regard its lack of presence as an indication that something was wrong with their rival.

Competitive trials were regarded as important because tractors and equipment could be seen working and their performance compared with rivals. The annual Long Sutton Tractors at Work Demonstration in Lincolnshire went on throughout the 1970s and, when technical problems arose at one of the first attended by the company during an early stage of the proceedings, there was an internal inquest of some severity because of the embarrassment caused in front of not just potential customers but also rival manufacturers.

A 272 awaits its turn at the Long Sutton Tractors at Work Demonstration. These comparative tests were considered to be of some importance because they could be used to gauge the performance of individual tractors against those of competitors.

A Leyland tractor during a field trial. Note the considerable number of front end weights.

An 85 hp tractor introduced in the early 1970s operating what appears to be some form of direct cultivation equipment during a field trial. The substantial power available from the turbocharged engine would be needed to operate such equipment.

The introduction of the 344 and 384 range took place at Alexandra Palace in north London. A stunt was staged whereby a new tractor was driven through a paper brick wall. This involved the demonstrator, Brian Webb, driving the tractor through a pre-formed opening covered with paper. Part of the framework was very close to the tractor and could easily have fouled its progress and caused a somewhat spectacular accident.

In 1969, at the time of the introduction of the 384 and 344 models, the endurance trial took place at Ratho in Midlothian, Scotland. It was then used as means of major promotion subsequent to the introduction of the new models.

After several years of rigorous rig and farm testing of prototype tractors fitted with the new Synchro gearbox, production was put in hand in 1978 and Turner decided to make a formal presentation at a major launch, held in Torquay. On 21 February, a presentation was made to worldwide distributors and that was followed on 27 February with a UK salesman's field day launch. The following day was a UK distributors' sales conference. These events both took place at the Imperial Hotel in Torquay, where ample indoor facilities were available for promotion to a large audience. In addition, a practical demonstration was to take place in a circus marquee on farmland north of

A rather more serious demonstration involved tractors fitted with the first of the Synchro gearboxes in the late 1970s. This 272 is fitted with a three-furrow mounted plough and is in front of the big top located north of Torquay.

Torquay. This location was chosen because South Devon was generally regarded as being reasonably free from the threat of snow and other extreme weather conditions.

A large 'big top' marquee was hired from Billy Smart's Circus and erected on a farm on high ground at the back of the town. It was intended that tractors would be displayed within and a presentation given to delegates, who comprised distributors and dealers from around the country. However, the planning came to nothing because this part of Devon had its heaviest snowfall for a considerable period of time. The marquee became unstable and riggers had to attempt to remove as much snow as possible from the roof.

In the event, it was decided to relocate that part of the promotion to the Imperial Hotel and hurried arrangements had to be made so the tractors could be displayed on the stage in the large ballroom. With the consent of the hotel management, doorways had to be enlarged so the tractors would fit through. The stage also had to be strengthened so it could support the tractors, which weighed several tonnes. A considerable amount of around-the-clock work was required to carry out these modifications but eventually all went smoothly.

The style and ambition of the presentation would have been very much the brainchild of Turner because it was typical of the sort of event that Ford had undertaken when promoting new tractor models. Turner was the host, making the principal introductions and providing the links between the speakers. The first presentation was given by marketing plans manager Tony Thomas, who was also in charge of product training. He outlined the history of the gearbox and gave the background to the company's development of the unit.

Thomas was followed by Bob Beresford, who was product engineering director, not just for tractors but in a more general role within the company's truck activity. Beresford had been much involved with tractor development and testing throughout the early part of the 1960s and had been promoted through a variety of roles which gave him access to both the truck and tractor parts of the business. His presentation was followed by Simon Evans, the supervisor for transmission designs who had been lead engineer in its development and in charge of the re-design of the gearbox. In addition, a slide presentation was given and despite the difficulties of the weather and the abandonment of the formal presentation within the marquee, the event went off satisfactorily.

Thomas made a final résumé of the three main presentations to provide a synopsis of the principal sales features and benefits that had already been described in some detail. He completed his talk by discussing the product training courses that would be available. In support of this, a mobile training school set up in a Leyland National bus was on show, already fitted out with a Synchro gearbox and the necessary service tools. This was subsequently to tour the country and was based at various distributors, where it was manned by John Patterson and Neil Spalding.

Within a day, the weather had cleared sufficiently for the practical demonstrations to take place where the marquee had been erected. At the demonstration site, Webb, Runciman and others provided practical demonstrations of cultivation work and the ability of the new gearbox to tackle smooth gear changes on the move while dealing with varying soil conditions and field gradients. The very wet conditions were particularly unpleasant for those involved with the demonstration, who had also suffered a lack of sleep because they had had to work long hours at the insistence of a senior member of the presentation scheme.

The crowning glory was the circus act put on by the team under the big top to show off the versatility and speed of the Synchro tractor in loader and handling work. This stole the show and brought tumultuous applause during every presentation through the week.

The journey to the event had been memorable for two of the more important participants. Evans accompanied Beresford on the trip from Scotland in his Rover car, finding the road conditions becoming increasingly difficult because of the heavy snow. At one stage it looked as though the route was going to become impassable because of broken down vehicles but, noting that the carriageway in the opposite direction was clear, Beresford drove over the central reservation and proceeded the wrong way down the other side of the dual carriageway to the next exit. There he joined the local road network and proceeded to the event.

The subsequent introduction of the Harvest Gold range was on a much more modest scale at the 1980 Smithfield Show, for reasons described in Chapter 9.

Export sales

In the 1960s, export sales were located at Cowley under the management of George Arnold, who had been in that role for some time. He was to retire in the mid-1960s after a decision was made to relocate the department to Longbridge.

The world was divided into six separate territorial areas, but some tractors were purchased by a government-backed department called the Crown Agents. This body acted on behalf of a number of former colonies and dependencies, on whose behalf they would purchase items such as tractors. Private

A 4/25 being demonstrated to a foreign purchaser, believed to be from Mexico.

sales went through distributors in the usual way. Australia differed because it had its own manufacturing facility at a plant on the outskirts of Sydney. Tractors were exported from the United Kingdom in CKD form and re-assembled at this plant with the local purchase of some components such as wheels and tyres being made as required.

Jeff Kitts had started with BMC in the mid-1960s and was initially allocated to Tractor Research, where he carried out a number of the pre-production tests on the Mini. Kitts moved to Far East Export Sales and was involved in introducing the Mini into Japan. This should have been a big market for the tractor but its lack of power meant it failed to achieve its anticipated sales. Local tractors, manufactured by Japanese companies such as Kubota and Iseki, were to become sufficiently successful for them to be exported worldwide, including into the UK, albeit some time after the Mini and its successor had ceased to be produced.

Kitts was instrumental in the export of 150 tractors

to Ceylon – now Sri Lanka. These sales were made through the local distributor Colettes, which dealt with demonstrations, dealer appointments and spares issues. Further Far Eastern agents were Werne Brothers in Malaysia, which also imported trucks, Aveling Barford rollers and JCB equipment.

European distributors included Domi of Denmark, JH Keller of Zurich, Switzerland, and Emile Frei in the same country. Vittorio Cantatore of Turin was a major importer, as it had been through the earliest Nuffield years. In the USA, tractors were exported through dealers including H Long in North Carolina, which was a significant machinery manufacturer and insisted on its name being shown prominently on the tractor bonnets. Other distributors included Millers of Wisconsin and there were others in Chicago, Pittsburgh and Oregon, although not all parts of North America had representation. In fact, the distributorships in the United States were relatively localised, as one would expect in such a populous country.

The company purchased Mosside Farm west of the Bathgate factory, and it was converted for a variety of purposes, including training.

Training school

BMC provided training for operators and salesmen and as long ago as the early 1950s this had taken place at the Tractor and Transmissions factory at Ward End in Birmingham. While tractor production moved away to Bathgate, other ancillary facilities such as design, testing, training and after-sales stayed and would only follow some years later. In the meantime, there was a

need to find new facilities for training and other functions, so the company acquired space at Atherstone airfield, near Stratford-upon-Avon. There was a substantial workshop facility available in an old hangar and a large area of ground around the runway for agricultural working.

A small training facility was set up under John Hiatt in portable accommodation. The workshop was

Some of the buildings used for training.

The principal training buildings were those in the bottom left of the picture; the farmhouse was used at least in part by security staff and the older farm buildings were used for storage. The demonstration bus parked in the yard was kept at Mosside when not in use in other locations.

available for demonstrating technical features and sales training took place in the modest portable classroom facility.

In 1969, Sir Donald Stokes's edict about moving ancillary parts of the tractor operation to Bathgate was to be fulfilled when the company relocated its training facility to Mosside Farm, a little to the west of the Bathgate plant. About 280 acres of farmland were available, of which 180 were farmed by a tenant farmer and the remaining 100 were available for the training and testing facility. Before that however, the portable classroom was moved from Atherstone to the end of C block at Bathgate as a temporary measure.

The farm buildings were renovated and the former cowshed became a separate workshop with an adjacent store in which some interesting items were kept,

The small reception at Mosside Farm. The tractor on the left is actually a half-sectioned 154 which was fitted against the blank wall as an unusual form of decoration.

including one of the prototype Standard tractors which had been acquired through the merger with Leyland the previous year. The farmhouse was converted for security staff and the front farm building was turned into a reception area and administration office with classrooms, in which a training supervisor and two lecturers provided tuition. This encompassed courses in sales and service for distributors and dealers but training for customers was available through the dealerships on a local basis. The courses were approved by a national training body and were subsidised financially, which made them popular.

The half-sectioned 154 survives in a collector's museum. This view of the reverse side shows the neat way in which the longitudinal cut was made.

The farmland at Mosside was very boggy and difficult to farm and this was in part carried out by tractor staff. The poor ground contributed to foundation problems in the buildings.

In the late 1970s, John Patterson ran the school with technical and, later, sales courses being provided. Later again the separate courses were merged but within ten years of its opening a further change of location took place following which Mosside Farm was taken out of use and subsequently demolished. That was not before

land at the rear was taken for the purpose of extending manufacturing facilities and this was to become the possible site of a new truck cab manufacturing facility, which would have been removed from the plant.

In 1980, a move was made to Beechfield House in Leyland and this appears to have taken place at about the same time the sales organisation was moved to nearby Preston. The new site offered training rooms, a separate workshop facility and 25 acres for field work. It was used for a limited time until the end of the tractor business early in 1982.

The entrance to the sales and service training facility.

Implements

In the post-war era, the development of tractor hydraulic systems led tractor manufacturers to design and make implements. In Ferguson's case, the majority of these tools were produced by external manufacturers. Ford quickly realised it was easier to enter into similar arrangements, although for nearly ten years the company manufactured a limited range of implements, principally ploughs and other tillage equipment.

When the Nuffield tractor was introduced at the end of 1948, it had been intended to manufacture implements as well. However, there was simply insufficient time and designers for such purposes and the company sought arrangements with outside manufacturers whereby it would 'approve' various equipment. Nuffield produced a small badge to that effect which was stuck to the implements and led to the mistaken inference they had actually been manufactured by the company itself. This misconception was reinforced by the fact a number of the implements were painted in a similar orange colour to the tractors.

The approved arrangement came to an end in the 1950s and as the decade wore on more and more implement manufacturers began to make equipment specific to particular models. They produced these in conjunction with the tractor designers to ensure they were adapted for connection to the hydraulic lifts.

When the Mini started to be developed in the early 1960s there were very few manufacturers with implements suitable for attachment to such a small tractor. Charles Black had persuaded Midland Industries to produce a prototype loader to go with the new Mini and Howards was requested similarly to introduce a rotavator for the same purpose. By the time the Mini was introduced there were high hopes

At first the marketing department thought that the Mini would be used exclusively for agriculture.

A considerable number of manufacturers produced equipment for the Mini and given that its sales were a disappointment one wonders how many of them actually made money. Here a set of Bamford harrows is being used.

A fixed-tine cultivator in use.

that it would be a major success and a considerable number of other manufacturers produced suitable implements. When a large-scale field demonstration took place at Alcott in north Warwickshire in 1966, areas within the estate were laid out to show the tractor working with a wide range of implements for all normal agricultural purposes such as grassland, arable and farm maintenance.

Among these were forklifts, which were not at that time seen as devices with a great deal of practical use on the farm. However, the company believed the small size of the tractor would, with appropriate adaptation and the use of Calor gas as fuel, make it suitable for working in buildings, other than those on a farm. The Mini could

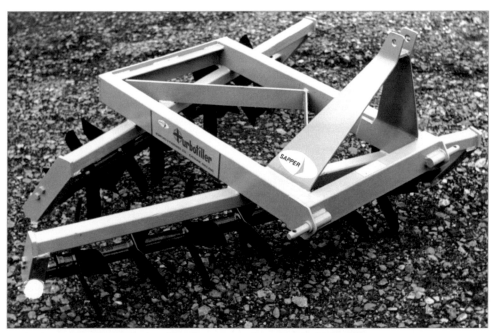

An item of cultivating/sub-soil equipment marketed as the Turbo Tiller.

also be used for moving stock and elevating materials into position within buildings with use of the telehandling equipment that started to be developed at this time.

Further sophistication came in the form of a front-mounted implement with rear cultivator, known as the Webb front-mounted hoe.

Midland Industries Limited produced a loader and this was first manufactured during the prototype testing phase. This 4/25 is being used for dung loading, driven by a youthful Ron Kettle.

One of the recruits to Tractor Research in the mid-1960s was a Canadian called Stan Whittingham-Jones. He started work demonstrating the Mini and mentioned he thought it would be ideal for use with grounds-care equipment. At the time this was a concept little known in the UK and fellow members of Tractor Research and BMC expressed some surprise. In North America, however, mechanically-operated grounds-care equipment was used extensively on playing fields, golf courses and similar places, where specialist equipment was

A Bamford two-furrow plough being used with the Mini. The substantial size of the components indicate this is an adaptation of what would have been a three- or four-furrow plough used usually with tractors of considerably larger size.

A McConnel mounted hedge-cutter in use.

used for rolling, brushing, aerating, seeding and generally maintaining large areas of amenity and sporting space.

Whittingham-Jones started to make enquiries at British firms to see if they had equipment that could be adapted. An arrangement was made with Essex company Sisis, which introduced a range of machinery for scarifying, rolling, aerating, brushing, spraying and other similar tasks. Whittingham-Jones also started negotiations with Dunlop, which was the predominant supplier of vehicle tyres to BMC, and it produced a specialised version which became known as a grassland tyre. This had less impact and so caused little damage to the surface upon which the tractor was operating.

A powered cultivator behind a Leyland 154. This would have absorbed a considerable amount of engine power.

BMC MINI TRACTOR

and maintenance equipment for sports turf, running tracks and artificial surfaces.

The BMC MINI TRACTOR is the right weight and power for every job on turf and track. The 948 c.c. BMC diesel cuts operating costs.

The BMC MINI TRACTOR with 'Hydraulics' is strongly recommended for all-round versatility and time-saving performance on this work.

BMC MINI TRACTOR with 'SISIS' 6-ft. Mounted Twin-drum Roller Ref. LR6
'SISIS' Brochure LE

BMC MINI TRACTOR with 'SISIS' 6-ft. General-purpose Aerator Ref. LA6
'SISIS' Brochure LE

BMC MINI TRACTOR with 'SISIS' 12-ft. Rake Scarifier Ref. LS/12
'SISIS' Brochure LE

Although the Mini had been manufactured with agriculture in mind, its biggest use proved to be in such areas as sports ground maintenance and highway landscaping. A small brochure was produced by the Sisis company showing equipment which could be used for such purposes.

Some of the earliest equipment produced, including a roller and an aerator.

A Mini fitted with a Lambourn cab towing gang mowers on a golf course. (©BMIHT)

The highway/municipal Mini was painted in a bright yellow livery and fitted originally with a number of separate items, including grassland tyres and square section mudguards to facilitate the use of a cab. It was considered that most local authorities would order cabs to provide comfort for their workers.

Among the range of highway implements was this substantial roller powered by a drive taken from the end of the rear wheel hubs.

A further highway fitment was this large leaf/debris collector, the Sisis Litamisa.

Among appliances that were thought suitable for use in industry was this substantial Hipope forklift. Given the size of the tractor one would have thought this could be a potentially hazardous item of equipment to use unless additional weights were provided to counterbalance potential loads on the tines.

The litter collector at work showing the clean sward where leaves have been removed.

The Sisis aerator mounted on the back of a twin-rear-wheel tractor.

It was quickly found this was a potentially lucrative market for the Mini and local authorities and other larger-scale managers of external amenity space became important customers. TA Parker and Son, a small horticultural supply company in suburban Worcester Park in south-west London, quickly became a major dealer for the Mini to the extent the firm was often the preferred supplier at trade shows. Another significant sales outlet was the Burgess Group in the Midlands for similar reasons.

A Lambourn cab fitted to a Mini hauling a single-furrow reversible plough.

A cab manufactured by Lambourn was also made available for the Mini, although subsequently Winsam produced a more robust metal and glass enclosure which had additional safety features to protect the driver in the event of a rollover and complied with post-1970 legislation. A safety cage had been available for the Mini if required.

Other specialist users of the tractor provided adaptations, including at least two that produced what effectively was a 'mini' JCB with a back excavator and bucket fitted. This was applied to the larger-capacity 25 hp tractor introduced in 1967.

The contemporary 4/65 and 3/45 tractors of larger size were, however, not provided with dedicated implements and the later Leyland ranges were able to use

A Winsam cab of early design mounted on a Mini.

Among the more substantial conversions was this front loader with a backactor seen here presumably under trial.

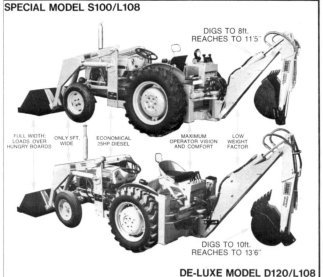

THE LIGHTWEIGHT CHAMPIONS
BY **INMAC**

COMPACT COMBINATIONS ······· RUGGED WORKHORSES
HIGHLY MANOEUVRABLE ············· REAL PROFITMAKERS

SPECIAL MODEL S100/L108

DIGS TO 8ft.
REACHES TO 11'5"

FULL WIDTH, LOADS OVER HUNGRY BOARDS · ONLY 5FT. WIDE · ECONOMICAL 25HP DIESEL · MAXIMUM OPERATOR VISION AND COMFORT · LOW WEIGHT FACTOR

DIGS TO 10ft.
REACHES TO 13'6"

DE-LUXE MODEL D120/L108

THE RIGHT MACHINES FOR
*Plumbers *Drainers *Concretors *Builders *Hiring Contractors
*Swimming Pool Excavators *Landscape Gardeners
*Municipal & Shire Councils *Electricity Authorities

Specifications Overleaf

implements from a wide range of suppliers. Bamford ploughs were often used in demonstrations and these were based upon the Scandinavian Kverneland range. The reason for this connection is that former Bamford employees had been taken on to work for the tractor business and this situation had also occurred in reverse, particularly when functions of the company were transferred to Scotland and former tractor personnel in the Midlands looked for work elsewhere in the area.

The merged British Leyland had within its myriad of subsidiary businesses a company called Barfords, which had a link to Aveling-Barford, a large-scale manufacturer of rollers, dumpers and associated civil engineering equipment. Barfords had a separate factory in Belton, near Aveling-Barford's plant at Grantham and produced a small range of implements including a reversible plough, drill, seed drill and grass topper. These were sold separately from the

Overseas manufacturers also used the Mini as the basis for small excavating equipment. This is a brochure for a digger made by the Australian company, Inmac.

A more significant adaptation is this digger with a small hydraulic crane manufactured by Long in the United States.

Larger tractors were adapted similarly and this is believed to be a 384 fitted with loading and excavating equipment and manufactured by a Scandinavian company.

The MIL loader was probably the most popular sold in the UK. It is seen fitted to a tractor with agricultural tyres which could be used for a variety of operations.

tractor business under the Barfords name. As part of the merger with Leyland, some of this equipment went on to be sold under the Leyland name and used the Bathgate address, although few of the tractor personnel with whom I have come into contact claim much knowledge of the activities of this company.

When the tractor business was sold in 1981, presumably the implement business continued but without the benefit of association with the larger group until the main Aveling-Barford business was sold.

Bomford produced a dozer blade for the larger tractors, as seen here in a brochure produced jointly for the tractor and implement manufacturers.

These Cultivator Frames are manufactured from 5 inch square section and can be fitted with double leaf type 'C' tines or double coil spring tines. Frames are designed for either cat. I or II linkage.

Models available

8' 6" Frame   9 Leaf Spring Tines   771 lbs. (349 kg.)
(2590 mm)

8' 6" Frame   13 Double Coil Spring Tines   788 lbs. (357 kg.)
(2590 mm)

Please Note!
This literature is only descriptive of a basic model(s). Specification of machinery can be varied to meet the individual requirements of customers. The weights given in this literature should only read as approximate. Furthermore the policy of the Company is one of continuous change and improvement and therefore our sales staff will be pleased to advise you of current information. All machinery is sold subject to the Company's Conditions of Business.

Leyland Tractors

TRACTOR OPERATIONS
LEYLAND SCOTLAND
BATHGATE
WEST LOTHIAN—SCOTLAND

Within the British Leyland companies there was a firm called Barfords of Belton, in Lincolnshire, which produced a small range of implements. These were the subject of this separate brochure, although it seems likely their manufacture was on a small scale. It is likely Leyland did not want to promote its own implements too much to avoid offending the larger implement manufacturers whose products were used regularly with their tractors.

Among other items of equipment produced by Barfords was the bale sledge.

Among cultivation equipment used was this two-furrow reversible plough, reputedly based upon a design from Belgium.

LEYLAND
DISC PLOUGH

Leyland Tractors

A three-disc plough of robust size, again marketed under the Leyland name.

time and time again...

sow accurately
grow profitably

BARFORD Combine Drills

The Barford combination drill was one implement that appears to have been manufactured rather more extensively.

LEYLAND
FLAIL CUTTER

This Flail Cutter is ideal for trimming the grass on highways, motorways, sports fields, golf courses, airfields, parklands, in fact anywhere where grass grows and is accessible for cutting.

It will also tackle rough cut, removing stubble, brush, scrub etc. to ground level.

For the farmer it can be applied for grass topping as height of cut can be adjusted from $\frac{1}{2}$" (12.7 mm) to 7" (178 mm).

Whilst the standard unit is supplied in an offset position, it can also be made available centrally mounted.

The mower is constructed of heavy gauge steel, overall width of the frame being 6' 8" (2032 mm)—actual cutting width 5' 7" (1702 mm)—and is fitted with 84 pairs of flail cutters mounted to a solid steel cutter shaft running in sealed bearings. The drive from the PTO is through a right-angle oil-lubricated gearbox fitted with heavy-duty gears and bearings, with a triple Vee belt drive from the lay shaft to the cutter shaft.

A heavy-duty safety flap covers the outlet of the machine and is adjustable to prevent solid objects being ejected when operating on highways or motorways. On the other hand the flap can be raised to permit a clear discharge when the Flail Cutter is used for grass topping. Height of cut is regulated by adjustment to the tubular steel roller, this being likewise mounted on self-aligning grease-packed bearings.

The three point linkage is designed to suit most makes of tractor fitted with either cat. I or II link pin and p.t.o. with overrun clutch.

Please Note!

This literature is only descriptive of a basic model(s). Specification of machinery can be varied to meet the individual requirements of customers. The weights given in this literature should only read as approximate. Furthermore the policy of the Company is one of continuous change and improvement and therefore our sales staff will be pleased to advise you of current information. All machinery is sold subject to the Company's Conditions of Business.

Leyland Tractors

TRACTOR OPERATIONS
LEYLAND SCOTLAND
BATHGATE
WEST LOTHIAN—SCOTLAND

The same firm produced a flail cutter, seen here in use at the side of a public road.

"Quick-fit" drive-in loader brings unbeaten versatility to the 502, 602 & 702 tractors

The QF Loader is entirely compatible with the new Leyland Tractors range, being purpose designed to give the utmost versatility to all middleweight two-wheel drive models. It is quickly and easily attached and detached on the drive in – drive out principle and provides a complete and really efficient unit for loading.

A wide range of attachments are available for the QF, which is produced to three alternative specifications, Standard, Extra Power and Hi-lift Extra Power. All are fitted with single acting cylinders on the lift frames with double acting cylinders controlling the bucket. A jib extension can also be fitted to the 602 and 702.

Safe, rigid parking is provided by a built-in stand that stows neatly into the basic frame. The jib extension is easily fitted, and there's a choice of lift rams.

The QF system offers a wide range of implements, including two manure forks, sludge plate, root crop bucket, grain scoop, three earth bucks, fork lift, silage grab, pallet fork tines, and a dozer blade.

More frequently seen in use were the loaders of different manufacturers, one of which is seen here working in conjunction with a Harvest Gold tractor.

BRISTOL

TRACTOR MOUNTED AIR COMPRESSOR

MODEL BL140

Free Air Delivery
140 C.F.M.
Guaranteed to BSS 726

- BRISTOL DUPLEX AIRFLOW COOLED TWO-STAGE COMPRESSOR

- FAST MOUNTING COMPRESSOR PACKS FOR ''255'' & ''270'' TRACTORS

- TESTED AND APPROVED BY BRITISH LEYLAND

- DOUBLE ACTING FOR VIBRATION FREE RUNNING

- AMPLE CAPACITY FOR TWO HEAVY DUTY BREAKERS

- HIGH GROUND CLEARANCE

- TOWING FACILITIES WITH COMPRESSOR MOUNTED

- SPARES AND SERVICE NETWORK THROUGHOUT UNITED KINGDOM

Air-compressor manufacturer Bristol produced a tractor-mounted model capable of being used with highway versions of the vehicle. This is one such model mounted at the rear of a standard tractor. The considerable additional weight at the rear must have required significant counterbalancing to make the tractor driveable on the highway.

CHAPTER 9

Finale

As can be seen in Appendix 2, the number of tractors manufactured at Bathgate was at its peak in 1973-5 but dropped sharply by 1980-1. During the period from 1972 until 1976, numbers fluctuated from between just over 14,500 to in excess of 18,700 units. By 1979, this had dropped to just over 11,250 and this was in the year when the highly successful Synchro gearbox became available. By 1980, fewer than half that number was built – just over 5,500 – and there was to be a further reduction in 1981 at the time of the introduction of the Harvest Gold range.

That same year, 1981, was to see Leyland begin to consider disposing of the tractor operation and meetings were held with potentially interested parties.

What is striking about the production figures is the difference between the number of manufactured tractors and skid units (see Appendix 3). The latter were produced principally for JCB but also for other users such as Coventry Climax, which adapted the engine and transmission for use in forklift trucks and other machines. This business was of considerable benefit to Leyland because at times when tractor sales were poor it was able to continue manufacturing the skid units in the knowledge there was a ready market which was less fickle than activities relating to agriculture.

At the same time as sales declined, the country was facing a very poor economic situation, which had afflicted motor manufacturing and British Leyland in particular. The company had got into such serious difficulties that by the end of the 1970s it was nationalised and under government control.

The Labour government had commissioned a report in the late 1970s from Sir Don Ryder which had suggested rationalisation of the company's activities and hinted at the disposal of peripheral activities, of which tractors would have been a prime example. This activity, together with the other large number of specialised vehicle manufacturing activities, was not regarded as being of long-term benefit to the business.

The work in updating the tractor has been discussed in detail elsewhere and comment made about the constraints on capital expenditure are answered by the larger issues facing British Leyland. Nevertheless, improvements to the tractor were beneficial because they maximised the marketing opportunities but also because they produced an up-to-date product which would have been attractive to a potential purchaser or as part of a merger with another manufacturer.

The possibility of merger had been considered seriously by the late 1970s and this might have saved the tractor business in some form. However, it can be safely assumed that, as with many other 'mergers' of a similar sort, there would eventually have been an effective takeover by the merging company. In the absence of a likely suitor with whom to merge, however, a sale became the only option by which to dispose of the business. Closing the business and liquidating the assets, such as they were, would have made a minimal return and there would have been the enormous costs of making redundant a large number of manufacturing and assembly staff.

It needs to be borne in mind that the production of the engines in particular was of benefit to the group as a whole. Supplying thousands of engines for tractors kept the unit costs down, which lowered the manufacturing cost of the trucks. Consequently, if the business had closed this benefit would not have been available.

In 1981 the company had produced a further business plan and this envisaged four main product ranges. These included a commercial division producing trucks and buses but, interestingly, there was no mention of tractors. That plan soon had to be completely re-thought because truck sales had fallen and it was reluctantly decided that, in order to produce further economies, one of the outdated truck ranges would be taken off the market and the tractor business sold.

That year, the company was considering an approach from Marshall of Gainsborough which was thought to be the most attractive option. This was mainly because this company had purchased Aveling Marshall, which produced the Marshall tractor used in agriculture and civil engineering. The business was owned by Charles Nickerson, a wealthy individual who had been introduced to the company by a Mr J Smart, a former deputy managing director of Leyland at Bathgate who had retired in 1979. On 15 December 1981, the company approved the sale to Marshall. There had been a separate approach from other interests, one of which was supported by managers within Leyland, but in the event the deal went through and the contract was completed in February 1982.

The domestic business was sold to Marshall and included within the disposal was the plant, parts and machinery and manufactured stock. This included about 600 tractors already made under the Leyland name and it had been agreed with Marshall that it would sell these tractors badged with the former's company name. This enabled Marshall to have stock available before it set up its own manufacturing line in the Gainsborough workshops.

This is reputedly the record card of the last Bathgate tractor manufactured.

The move of the training school to Lancashire in the 1970s resulted in the Mosside Farm complex becoming derelict. One assumes the somewhat remote location made it uneconomic to try to sell the buildings.

A line of tractors leaving Bathgate on their way to Gainsborough in Lincolnshire, where they were to be sold under the new arrangement with Marshall.

This, however, was not quite the end of British Leyland tractors because at a depot at Oldbury in Worcestershire a batch of in excess of 30 machines had been prepared for sale to Zaire, Marshall having only bought the domestic tractor business. These tractors had been in store for some time and had remained in their original livery when the Harvest Gold range was introduced. They had deteriorated somewhat through lack of use and exposure to the elements and consequently had to be serviced fully, repaired and repainted to make them suitable for sale.

The disposal of the business was not without its critics, principal among whom were a number of trade union members at the factory and politicians. Consequently, an investigation was made by a House of Commons committee into a number of allegations, among which were that the profitable business had been disposed of for a knock-down price. There were also allegations of impropriety regarding the sale by various persons connected with Leyland and criticism

of late decisions by the company regarding investment, which it was alleged benefited a purchaser rather than the firm itself.

In addition, allegations were made regarding the frequent management changes some of which were associated with the production of various plans for reorganisation of the company. Some of these changes were considerable and involved the acquisition of additional management expertise on various levels which, from the benefit of a more distant and objective examination, seem remarkable given the parlous state of the business. It is little wonder those on the shop floor thought there was something untoward in the arrangements being made. It has to be said in Leyland's defence that its reorganisation enabled the four main sectors of the business to trade on an individual basis and, while this may at the time have seemed to have been somewhat wasteful of resources, it nevertheless provided the framework for either the potential sale on a larger scale, which effectively happened in the mid-

A tractor on display at a vintage vehicle event showing the standard of finish that many collectors achieve and which is often well in excess of the standard that would have been produced for normal commercial purposes.

1980s, or its continuation with the addition of external private finance.

In the end, British Leyland identified that its tractor business had failed because:

1. Its volume base was too low to achieve competitive delivered costs against high-volume international competitors.
2. The facilities employed to make tractors were excessive in relation to the volumes actually built and this contributed to the uncompetitive cost structure.
3. Overall demand for tractors had been depressed, a feature that would have applied to other manufacturers also.
4. Profit margins on export tractors were being reduced or eliminated by the appreciation of sterling against other foreign currencies, greatly favouring tractors being imported from competitor countries where currency fluctuations had lowered their potential costs.

In addition, actions undertaken by management to sustain the business, including the introduction of the Golden Harvest range, were not sufficient to overcome the fundamental structural problems of low overall demand, lack of competitive costing and the high exchange rate. These reasons, of course, are reflected in the figures in the appendices and speak in black and white terms of the reasons why the business was ultimately to go the way it did.

Thus ended a business set up in the optimism of the post-war era to produce a well-designed and attractive machine that was technically as good as the best of the competition. It has to be said that the numbers produced since 1948 were small compared with those made by Massey Ferguson and Ford, whose potential for profit was therefore always going to be greater when sales were good. That gave those companies the benefit of being able to generate higher internal profits, which could be ploughed back into the business to fund more advanced technical developments.

A Harvest Gold Synchro tractor being inspected by prospective purchasers at a Cambridgeshire auction. This vehicle could go to a collector or may see further agricultural use.

Consequently, the Leyland tractor business was always at a potential disadvantage because sales and profitability would always be lower than other mainstream activities of the Leyland Group, such as car manufacturing and commercials. In the dire situation in which it found itself in the late 1970s and beginning of the 1980s it is unlikely any other proprietor would have retained the business given the circumstances.

There is now only one major tractor manufacturing plant in the UK, at Basildon, but its original owner and builder, Ford, now has just a small holding in the business of a much larger agricultural manufacturing conglomerate. Indeed, its name does not even feature in the title of the company, Case New Holland. Massey Ferguson, one of Leyland's great rivals, no longer manufactures tractors at all in the UK, but is superficially less affected by change of the sort seen

elsewhere because it is part of a larger manufacturing conglomerate, as is the case with Ford.

If British Leyland had been able to survive the traumas of the 1970s and '80s, it would almost certainly have ended up in a similar way. Sadly, the subsequent failure of Marshall and others who subsequently bought the business, ensured that it survives now only in the ownership of one of its surviving dealerships, J Charnley and Son. The tractors it produced remain both in agricultural use and increasingly as preserved machines with collectors. The company's memory will be maintained in the great enthusiasm for the product of those involved in its production and sales, all of whom, one is sure, would have hoped that a better outcome for the business would have been achievable, rather than the sad way in which it was to end.

APPENDIX 1
Model Introduction Dates

Mini	1 Dec 1965
3/45 4/65	26 June 1967
4/25	22 November 1968
154 344 384	25 November 1969
253	December 1971
	(498 Engine October 1972)
245 258 279 (154) 285 485 2100 4100	29 November 1972
262 272	26 November 1972
Middleweight's Q-cab	May 1976
Synchro Gearbox	24 March 1978
462 472	3 July 1978
282 482 235	December 1980
Leyland Golden Harvest 302 502 602 604 702 704 802 804	December 1980
Leyland ⟶ Marshall	March 1982

APPENDIX 2
Bathgate Tractor Production 1963 – 81

1963	1964	1965	1966	1967	1968	1969	1970	1971	1972
3,743	13,953	13,998	17,770	13,963	14,175	13,591	14,077	10,736	14,697

1973	1974	1975	1976	1977	1978	1979	1980	1981
18,766	14,929	18,713	16,409	14,770	12,108	11,254	5,522	3,821

APPENDIX 3
Tractor Sales in the UK by Model 1969 – 79

	1969	1970	1971	1972	1973	1974	1975	1976	1977	1978	1979
Wheeled											
154	6	394	326	274	301	272	156	101	131	116	68
245	-	-	2	800	464	434	251	214	192	143	98
262							1	580	402	373	291
272							1	864	802	681	802
285					58	86	96	116	77	20	74
2100					58	146	58	79	44	11	5
462										12	25
472										43	123
485							1	-	1	-	-
4100							4	3	2	-	-
282											27
482											5
345/344/255	195	1,040	691	753	808	774	456	76			
465/384/270	1,278	1,615	964	1,055	1,307	1,446	1,078	101			
425/916	501	52									
Total Wheeled	1,980	3,101	1,983	2,882	2,996	2,858	2,102	2,134	1,651	1,399	1,518
Skidded											
262								1	2	-	
272								23	98	1,871	2,177
472											28
345/344/255	430	132	-	-	5	1	9	9			
465/384/270	3,375	3,570	3,070	5,304	5,977	4,634	3,853	4,762	4,161	4,368	2,627
Total Skidded	3,805	3,702	3,070	5,304	5,982	4,635	3,592	4,795	4,261	6,239	4,932
Total Tractors	5,785	6,803	5,053	8,186	8,978	7,493	5,694	6,929	5,912	7,638	6,350

APPENDIX 4

Tractor Exports by Model 1969 – 79

	1969	1970	1971	1972	1973	1974	1975	1976	1977	1978	1979
Wheeled											
154	3	434	725	2,094	3,765	3,145	5,376	4,420	2,970	961	500
245				559	613	439	695	419	679	577	556
262								874	1,258	809	651
272								2,284	3,290	1,938	1,675
285					12	8	65	406	346	23	23
2100					8	25	125	196	121	38	22
462										8	73
472										62	302
485									3	-	-
4100								1	4	-	1
184											990
345/344/255	919	1,906	1,370	902	1,337	1,017	1,329	199			
465/384/270	4,934	4,760	3,543	2,857	3,881	2,669	5,294	591	7		
425/916	114	46									
Total Wheeled	7,000	7,146	5,638	6,412	9,616	7,333	12,884	9,390	8,678	4,417	4,793
Skidded											
262								8			
272								12	180	51	110
472											1
345/344/255	76	65	-	-	-	-	11				
465/384/270	174	63	45	99	172	103	124	70	-	2	
Total Skidded	250	128	45	99	172	103	135	90	180	53	111
Total Tractors	7,250	7,274	5,683	6,511	9,788	7,436	13,019	9,480	8,858	4,470	4,904

APPENDIX 5

Tractor Sales by Region 1969 – 81

	1969	1970	1971	1972	1973	1974	1975	1976	1977	1978	1979	1980	1981
Major Markets		2,504	2,914	2,188	3,015	2,177	3,449	1,695	1,856	1,383	1,456	500	351
Region A		371	387	238	208	110	157	45	43	109	56	97	102
Region D		347	161	1,589	2,925	2,687	5,041	4,517	3,252	681	1,109	1,858	1,131
Region E		724	635	406	504	521	825	299	511	147	114	75	116
Region F		738	248	241	872	509	952	626	501	91	175	123	71
Europe		2,227	1,194	1,728	2,062	1,313	2,463	2,008	2,422	1,549	1,715	1,076	612
Eire		332	130	121	202	115	132	290	273	510	279	35	40
Misc		31	14	-	-	4	-	-	-	-	-	-	
Total Exports	7,250	7,274	5,683	6,511	9,788	7,436	13,019	9,480	8,858	4,470	4,904	3,764	2,423
Total UK	5,785	5,803	5,053	8,186	8,978	7,493	5,694	6,929	5,912	7,638	6,350	1,758	
UK Excluding Skids	1,980	3,101	1,983	2,882	2,996	2,858	2,102	2,134	1,651	1,399	1,518	1,102	1,398
Skids	3,805	3,702	3,070	5,304	5,982	4,635	3,592	4,795	4,361	5,239	4,832	656	
Total Deliveries	13,035	14,077	10,736	14,697	18,766	14,929	18,713	16,409	14,770	12,108	11,254	5,522	3,821*
												* Exc SKIDS	
Exports BU/KD													
BU		7,263	5,578	4,893	6,874	5,290	8,909	5,335	6,570	4,242	3,880	2,114	1,062
KD		11	105	1,618	2,914	2,146	4,110	4,145	2,288	228	1,024	1,650	1,361

APPENDIX 6
Tractor Sales in Eire and Europe 1970 – 81

	1970	1971	1972	1973	1974	1975	1976	1977	1978	1979	1980	1981
Belgium	80	33	13	7	8	*	27	18	23	1	7	-
Denmark	578	118	326	589	303	273	317	358	183	315	21	123
Finland	982	359	844	690	373	991	870	1013	486	493	509	257
France	40	143	54	130	89	161	130	54	14	-	6	3
Germany W	2	-	-	-	-	-	-	-	-	-	-	-
Italy	20	31	3	12	7	-	3	16	22	222	166	6
Netherlands	44	13	62	71	40	91	54	162	121	77	25	6
Norway	179	319	136	239	51	132	129	191	61	78	50	77
Portugal	105	63	100	59	35	206	241	488	210	174	108	45
Sweden	168	91	156	88	125	178	186	66	112	99	32	50
Switzerland	7	22	-	22	23	13	1	4	-	-	-	-
Canary Is	12	2	9	2	4	-	-	-	-	-	-	-
Greece	-	-	-	50	74	356	50	52	317	256	152	45
Iceland	-	-	3	-	-	-	-	-	-	-	-	-
Spain	-	-	22	1	2	-	-	-	-	-	-	-
Yugoslavia	-	-	-	102	179	62	-	-	-	-	-	-
Total Europe	2,227	1,194	1,728	2,062	1,313	2,463	2,008	2,422	1,549	1,715	1,076	619
Total Eire	332	130	121	202	115	132	290	273	510	279	35	40
Misc	31	14	-	-	4	-	-	-	-	-	-	-
			53									

APPENDIX 7

Tractor Sales to Major Markets outside Europe 1970 – 81

	1970	1971	1972	1973	1974	1975	1976	1977	1978	1979	1980	1981
Australia	321	114	200	546	410	1,202	554	1,101	350	281	23	
Canada	416	401	177	362	143	324	85	93	59	99	46	
India	2	1	250	285	-	-	-	-	-	1	-	
South Africa	723	326	459	112	374	821	115	41	235	107	262	47
*USA	1,042	2,072	1,102	1,710	1,250	1,102	940	621	739	968	169	304
Total	2,504	2,914	2,188	3,015	2,177	3,449	1,695	1,856	1,386	1,456	500	351
*USA KD Content	10	103	238	374	-	-	-	-	34	-		
* Inc Canada												

INDEX

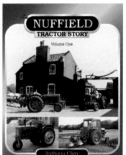

The Nuffield Tractor Story Volume One

Anthony Clare

Anthony Clare starts with the preparatory work of 1943 and takes the story up to 1967, the era of the powerful 10/60 model. He deals fully with development, production, models, sales, distribution, implements, after-sales service and testing.

Hardback, 192 pages inc. 303 photos.

Marshall Diesel Tractors

Peter Anderson

In this highly illustrated book Peter deals with: the 'Colonial'; Model E (15/30); 18/30; 12/20 and M; Field Marshall Mark 1 and Series 2, 3 and 3A; MP4 and MP6; Fowler VF and VFA; Track Marshall, and Gainsborough loading shovel.

Hardback, 220 pages inc. 280 illus.

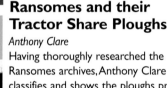

Ransomes and their Tractor Share Ploughs

Anthony Clare

Having thoroughly researched the Ransomes archives, Anthony Clare classifies and shows the ploughs produced by the company. He deals with the horse ploughs adapted for tractors, mounted ploughs, Ford-Ransomes, the links with Dowdeswells and also includes analyses of identification codes and the TS classification.

Hardback book, 100 pages inc. 120 photos.

Farming in Miniature Volume 1: Airfix to Denzil Skinner

Robert Newson, Peter Wade-Martins & Adrian Little

The first of two comprehensive volumes reviewing British-made toy farm models up to 1980.

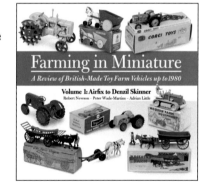

Volume 1 contains nearly a thousand captioned photographs and clear introductions to seventeen manufacturers/brands including Britains, Corgi, Chad Valley, Crescent and Charbens.

Hardback book, 360 large format pages inc. 970 photos.

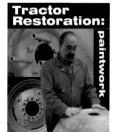

Tractor Restoration: Paintwork

Alan Davies

This two-DVD set shows all the steps in painting a tractor at home: dis-assembly; repairing light damage; filling, sanding and stopping; priming and finishing.

2-DVD set, 142 minutes.

The MF35 Workshop Service Manual

Chris Jaworski

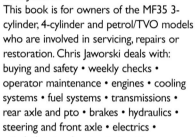

This book is for owners of the MF35 3-cylinder, 4-cylinder and petrol/TVO models who are involved in servicing, repairs or restoration. Chris Jaworski deals with: buying and safety • weekly checks • operator maintenance • engines • cooling systems • fuel systems • transmissions • rear axle and pto • brakes • hydraulics • steering and front axle • electrics • bodywork. In each chapter he includes step-by-step instructions illustrated with clear photographs – over 650 in all.

Hardback, 208 pages inc. 650 photos.

International at Doncaster: British production 1949–75

Stuart Gibbard

The 27 machines include examples of the larger wheeled tractors produced at Doncaster as well as the smaller ones from the Bradford plant. Tracked models are covered in detail. The tractors shown are all currently working or restored.

DVD, 85 minutes.

One Farm, One Year

Chris Lockwood

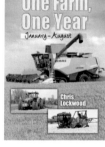

Two DVDs which together constitute the most detailed programme ever released about the farming year on a large arable enterprise. Loose Hall Farm Ltd runs to 3,600 acres on eight farms stretching over eight miles – large enough to require serious project management and run some excitingly large tractors and machinery.

Two-DVD set approx 4 hours.

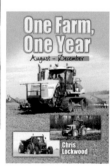

Old Pond Publishing Ltd, Dencora Business Centre, 36 White House Road, Ipswich IP1 5LT, United Kingdom.

Phone +44 (0) 1473 238200 • *Full free catalogue available* • **www.oldpond.com**

ABOUT THE AUTHOR

Tony Clare is a Chartered Building Surveyor from Surrey whose lengthy vacations on a family farm near Salisbury in Wiltshire inspired a lifelong interest in agricultural history and in particular machinery. Among his collection of vintage agricultural equipment is a Nuffield M4 which he has owned for ten years. Tony maintains the M4 in full working order and regularly competes with it in south-eastern ploughing matches.

The Nuffiled Tractor Story Volumes One and Two is the second project Tony has written on an agricultural history topic, although he has undertaken private research on various subjects over the years. His previous book is *Ransomes and their Tractor Share Ploughs*, published by Old Pond. It is Tony's long-term interest in Morris and its agricultural division that has been the spur to researching and writing this history of the Nuffield and Leyland tractor.

The Nuffield and Leyland Tractor Club

The Nuffield and Leyland Tractor Club promotes the interests of the owners of Nuffield, Leyland and Golden Harvest Marshall tractors. Their information-packed website is at www.thenuffieldandleylandtractorclub.co.uk.
Other points of contact are:
Mrs Pam Harlow, Bucketwell, Frocester, Glos GL10 3TG.
Email pam.harlow5@btinternet.com.
Phone +44 (0)1453 828737

J. Charnley & Sons

J. Charnley and Sons are specialist suppliers of Nuffield, Leyland and BMC parts. Their website is www.charnleys.com.
Other points of contact are:
J. Charnley & Sons, Marsh Lane, Brindle, Chorley, Lancs PR6 8NY.
Phone +44 (0) 1245 854103.